Studies in Fuzziness and Soft Computing

Volume 354

Series editor

Janusz Kacprzyk, Polish Academy of Sciences, Warsaw, Poland
e-mail: kacprzyk@ibspan.waw.pl

About this Series

The series "Studies in Fuzziness and Soft Computing" contains publications on various topics in the area of soft computing, which include fuzzy sets, rough sets, neural networks, evolutionary computation, probabilistic and evidential reasoning, multi-valued logic, and related fields. The publications within "Studies in Fuzziness and Soft Computing" are primarily monographs and edited volumes. They cover significant recent developments in the field, both of a foundational and applicable character. An important feature of the series is its short publication time and world-wide distribution. This permits a rapid and broad dissemination of research results.

More information about this series at http://www.springer.com/series/2941

Enric Trillas

On the Logos: A Naïve View on Ordinary Reasoning and Fuzzy Logic

Enric Trillas
University of Oviedo
Oviedo, Asturias
Spain

ISSN 1434-9922 ISSN 1860-0808 (electronic)
Studies in Fuzziness and Soft Computing
ISBN 978-3-319-85815-9 ISBN 978-3-319-56053-3 (eBook)
DOI 10.1007/978-3-319-56053-3

Softcover reprint of the hardcover 1st edition 2017

Printed on acid-free paper

This Springer imprint is published by Springer Nature
The registered company is Springer International Publishing AG
The registered company address is: Gewerbestrasse 11, 6330 Cham, Switzerland

Foreword

-In the beginning was the Logos (John's Gospel)-

Just looking at the title and observing that formulae constellate the pages without the overwhelming presence that has become customary in scientific writing, provides a refreshing sensation of going back to the Greek serenity hinted at by the title.

The title is, usually, an indication of what the author had in mind when he began to conceive of the book. So, let us start from it: *On the Logos*. Oh it is truly demanding! In the Prelude Enric explains that he refers to the "capabilities of reasoning and expressing in some plain language" that are "shared by all mentally sane people." We (and the reader should) go back to it after finishing reading the book.

Now we start from the subtitle and the concepts he presents in the first pages. Well, these concepts look more attainable. Ordinary reasoning is a very difficult and challenging theme and I know that it is a comprehensive argument for attainment for which Enric in the last years has used all the technical, formal, and conceptual instruments of his toolbox. It is what he has already done not only in many scholarly papers but also in two recent books, one in Catalan, the other in Spanish. But the present one is something different: it is neither a translation nor a reshuffling of them. It is really a new "synthesis". Synthesis of what? Of all his research, a summary of the path Enric has followed over the years, but also of the questions, the crucial questions that have accompanied the scientific revolutions of the twentieth century—in physics, in mathematics, in logic—and, among them, also in the newcomer to the club of scientific disciplines that can generically identify with "information sciences" but still lacks a stable name after having changed many over the years (from cybernetics to AI, cognitive sciences, etc.).

These were paths that Enric naturally crossed during his studies and his career. However, he did more: he searched and looked for unusual aspects, pushed by his intellectual curiosity and open-minded approach to these problems. The book spans many chapters dealing with or touching upon such different topics as "meaning", "thinking", "analogy", "(ordinary) reasoning", or "reasoning in quantum physics". Different topics but united and unified by the common aim of contributing to understanding them by using a scientific approach to natural (better, ordinary, or plain) language and reasoning, which is the pervasive thread of the book.

"Scientific" for Enric means—among many other things—"measurable". Also meaning should be measured. This measure, however, needs to come out of a correct adequate modeling of what we are studying, not out of a mechanical routine application of already available results, not the routine work of "applied mathematics" that superimposes known techniques and results, in a sort of Procrustean bed.

We must be aware, as Enric writes, that "plain languages are not the creation of a single person, but are slowly generated by linguistic interactions between groups of people; plain languages are socially constructed over the course of time, and words acquire meaning along such interactions." Clear! Is it also simple doing a model of this? Rhetorical question.

I am inclined to think that Enric does not believe in "applied" mathematics as a separate category, but in a "new math" for "new problems" (a new math *adequate* to the complexity and novelty of the problems). In a sense this is not customary today (many people use not only the expression "applied mathematics" but also the one of "applied mathematician"); it is natural, anyway, and above all, is not new at all. (Isn't that what Newton did when he invented the calculus for describing Nature?) Newton was not an applied mathematician.

The division of the book into two parts, one dedicated to "sowing" ideas, the other to "gathering" questions can also be seen as typical of Enric's style (human before the scientific, and—for this same reason—strongly scientific). We must propose and defend new ideas trying to convince everyone that these can be good points of departure for further investigations (fallible, of course, as is everything in science) but, at the same time, we must ask new questions, trying to discuss them openly involving in the project as many people and attitudes as possible in order to reach a shared common point from which to start (and later divide). Let me write that I saw with great interest that the problem of "(non) monotony in reasoning" has been put in this second part. We really should share a few common conceptual points before trying to confront the question of constructing quantitative models of this feature of reasoning, simple and clear models, not cumbersome machineries of which the scientific literature is full and which remind us of Ptolemy's epicycles.

So, after all, Enric was right in calling his book *On the Logos* inasmuch as not only, "In the beginning was the logos," but, as John goes on to write, "All things were made through the same, and without the same nothing was made that was made." The "same", as I read it, is the Logos. In addition to the Greek use of the term, we perhaps should also take into account the use by John in his Gospel. Are we dwelling outside our proper territory? I do not think so. Let me be clear at this point. Enric took the term seriously and the reader should take it seriously too. Maybe Enric has in mind only, or mainly, the Greek interpretation and would consider the present extension improper. I think, however, that every reader should take into account all the possible interpretations of this as well as all informal notions we encounter reading the book (the book is doing exactly this regarding ordinary reasoning).

What John did was to try to connect Greek concepts and philosophy with the "needs" of the new, innovative, religion. Today science and scientific knowledge and tradition play a crucial role in our society. We should be able to do similar difficult and profound conceptual operations when trying to scientifically understand and address difficult questions. The problem of the way in which the human being reasons, in general, not when he is concerned with specific and limited chapters of his action, is one such big question.

This is what Frege attempted to do, and Boole before him, the latter successfully but, seen from today's perspective, in a very limited range, and the former constructing a wonderful cathedral that, unfortunately, was based on very unstable ground. His project, his program, however, was so grandiose and magnificent that a lot of today's philosophy of language is still full of his ideas.

In this perspective, in our reading, the Logos is at the same time both the "Informal," the "Tao," the "Chaos," and what helps in putting order in it, the "Reason," the "Measure." In this second role we have here something very similar to what is alluded to in the title (and the ideas) of a book by Nelson Goodman, *Ways of Worldmaking*. In fact, trying to understand (pieces) of Nature, we are constructing "worlds," or, as scientists and mathematicians say more modestly, "models."

As Enric writes, "A model should come from some knowledge of the subject, and a good model allows finding novelties. In the case of ordinary reasoning, models should come from the current *scarce* knowledge of it" (my Italics).

A model of such a general thing not only is, obviously, very difficult to obtain, address, and study but presents additional problems (apparently unsurmountable) also of methodological and epistemological type, because we are anyway obliged to delimit what we want to study without losing essential features of what we want to understand. This theme periodically comes up in the book.

We cannot but congratulate Enric for this book that presents the first and basic bricks of what appears to be a very interesting and ambitious research project whose aim is one of attaining and solving one of the most challenging and difficult open problems, a project based on a few already assessed basic results and which starts from and encapsulates many findings of the research done by himself. An open project, however, that is open to the finding of nearby disciplines, mainly neurosciences, and usefully open to suggest crucial questions also to these cognate disciplines.

Without doubt many characters will appear on the stage and many of those already present will change their roles as soon as neurosciences provide us new relevant information. However, we dare to say that the overall structure of the book is general and robust enough to sustain and support many changes.

As it stands, the book seems to conclude the parable of the scientific quest done by Enric, not in the sense that he will not provide, in the future, other challenging contributions, but in the sense that in this book both the philosophical questions (which have always been at the core of his intellectual interests and curiosity) and his mathematical findings (that can give support, flesh, and substance to the disembodied ideas which are the most fantastic products of human creativity and imagination) are here fused together in an admirable way.

However, I am sure that the present synthesis on the Logos, on his thoughts on "Thought", are not his final word (his final logos). And we, while appreciating the way in which he has addressed—*now*—these difficult questions, remain waiting for the unexpected suggestions and comments that Enric will provide as soon as new findings ask for new paths to be followed.

Settimo Termini
University of Palermo (Italy)

Contents

Prelude: Guessing, Telling, and Computing

A new frontier seems to be opened for natural science thanks to the advances in neuroscience's knowledge concerning the human brain's functioning and, in particular, its derived natural phenomenon called thinking. Before such borders will be actually crossed, it cannot be foreseen how the essentially person-driven manifestation of thinking called ordinary, everyday, or commonsense reasoning, will be scientifically understood.

Thinking comprises much more than reasoning, and it is beyond doubt that ordinary reasoning, the spring of rationality, is mainly expressed in natural language's words and statements which, by consequence, can be seen as carriers of information; reasoning and language are in a deep semantic interlinking. Consequently, it seems reasonable to study ordinary reasoning pragmatically by not keeping it too far from its usual expression in natural language; natural language is acquired and fueled by ordinary reasoning, and reciprocally.

It should be pointed out that the expression "natural language" only has sense for distinguishing the plain everyday languages from the artificial ones built up in the formal sciences, but that, actually, a plain language is not properly something naturally fixed, but continuously evolving, influenced, and "loaded" by, for instance, the culture surrounding its speakers.

At the end, the capabilities of reasoning and of expressing themselves in some plain language, the old Greek's *Logos*, are two essential and intermingled characteristics of rationality shared by all mentally sane people on the Earth; at least, it is thanks to plain language that a person can know the reasoning made by other people. *Logos* was translated into Latin by *ratio*, not referring to a ratio of numbers but to a balance between what is pro and what is con in respect to some statements submitted to scrutiny. For this reason, it cannot seem too odd to try to accomplish a symbolic study of ordinary reasoning as it is done by people, through plain words and statements, just excluding contradictions but neither with the strong constraints imposed by "logical laws" (not always presumable in language), nor abusing any artificial language, but keeping the study as close as possible to the ways in which ordinary people seem to reason. That is, done in a mathematically naïve style.

For instance, the classical representation of the linguistic "and" by a lattice conjunction is abusive with respect to, at least, the presumptions of its commutative property, of being the greatest element below the statements the "and" joins, and always identifying "p and p" with "p". Formal representations such as a lattice's, impose on plain language and ordinary reasoning "laws" that often cannot be presumed and which, in any case, should be carefully tested before introducing them.

Because understanding what words actually denote is essential for reasoning, the study of ordinary reasoning requires scientifically domesticating the (dark) philosophical concept of meaning. As shown here, such domestication can be done in a form similar to how probability was domesticated, even if it were in the limited conceptual and strongly structured frames of Boolean algebras and orthomodular lattices. Inasmuch as meaning is attributed to words in any setting, such domestication should be done in the general and, in principle, mathematically unstructured frame of plain language, by giving extension to meaning, that is, measuring it; measuring is but a necessity and a characteristic of science. After doing this, it should be analyzed as to what ordinary reasoning consists in, the role analogy plays as a first engine of it, and the difference between deduction and induction clarified, that is, between the extremes of deductively deploying what is just hidden in the departure's information, or evidence, and reaching something new, or creative reasoning. Deduction and creation are the pearls of reasoning, gracing both the refuting of arguments and the abducing of hypotheses, as well as facilitating the proof of "theorems" in the deductive own formalism of mathematics. Because deducing in ordinary reasoning often leads to contradiction, it should be differentiated from deducing in a formal frame where reaching contradictions is never accepted.

In this respect there is a recent development remembering, up to some point, how the old scholastic logic did confront reasoning, but was concerned with a symbolic form of representing ordinary reasoning. A development that, based on language and very few and soft formal constraints, came directly from the conceptual skeleton of some fuzzy logic views, indirectly from the neopositivistic philosophical reflections made by the Vienna Circle's members, and also from some mathematical studies.

It is on these antecedents that what follows is based by following, in its argumentation, William of Occam's razor methodological rule, advising us not to introduce more entities than those strictly necessary, but also taking into account Karl Menger's balancing addenda of not introducing fewer entities than those allowing us to capture something new, previously unseen or unknown. It should be remarked that such rules refer here to the considered linguistic, or reasoning's operations, and the "laws" to which they are supposedly submitted for obtaining both a symbolic representation and a calculus able, at the end, to translate them into a formal framework.

At least and perhaps, such developments can be suggestive for those aiming at the Menger's "exact thinking," and wishing to reflect on the flexible, neither formal, nor always deductive, ordinary reasoning. It is done in what can be possible, by pragmatically proceeding in ways simpler than those currently covered under the label of "mathematical logic" that, usually, go on under an artificial language, rigid and strictly fixed by some finite number of axioms. A novelty in the developments is the classification of conjectures in consequences, hypotheses, and speculations. Once introduced, for the first time, this last particular and deductively wild third class of conjectures, formerly not previously differentiated in the literature, seems to produce, in association with analogy, the creative reasoning that, without previous knowledge, analogy, and speculation, seems to be impossible. A good enough example is the story of how the German chemist August Kekulé discovered the structure of the benzene molecule after an 1861 fantasy dream of a snake eating its own tail (the old *ouroboros* of alchemy), a discovery that later on, and thanks to the forthcoming theory of the chemical molecular structure, showed its great fertility by leading to the success of the German industry of colorants, a discovery that seems to be impossible without the previous knowledge Kekulé held on benzene's properties.

It is very difficult to arrive at something new without seeing a neighborhood of previous knowledge of the subject, and constituted by hypotheses, refutations, consequences, speculations, and analogies. Continuing Karl Popper's ideas, if scientific reasoning always consists in conjectures and refutations, it is also important to know how they are reached or abandoned; at the end, and at least in the scientific context of searching, analogy with former knowledge is basic.

This book just pretends to deal with a pragmatic view of common sense reasoning shared by both specialized and ordinary people, and in such a form that, at the end, can admit to being particularized (by adding more suppositions) into specialized modes of reasoning, such as they are classical and already known, models for several types of formal reasoning, namely the models for precise Boolean, quantum orthomodular, and the imprecise De Morgan or Zadeh, each one living in different rooms of the common house of ordinary reasoning.

At the end of each section and as recommended reading, a list of references appears. All of them influenced what is in the book, and are only suggestions to the reader that can help her or him to go deeper concerning the corresponding subjects. After the final section, a complementary list of books is added; without reading at least some of them, no good enough perspective on the ground subject of this book may be grasped.

Ordinary reasoning is neither, in the large, of a formal deductive character, nor is always done in a previously formalized setting. For its part, natural language is not deeply studied here in itself but taken, ideally, as it appears to be; apart from the essential domestication of the meaning of words and statements, only some basic topics of language leading to building statements with words, such as conditionality, connectives, and modifiers, are analyzed and, as much as possible, try to be mathematically modeled. In the same vein, the new model of the meaning of words allows us to clarify what a fuzzy set and its membership functions are and can, eventually, help to shed some light on the debate among the two basic interpretations of probability, the objective and the subjective, and for what concerns, at least, plain language's use of "probable", "possible", and "uncertain".

Even if, further on, the progress neuroscientists are making regarding the knowledge of thinking can force us to modify the views sustained here, the presented mathematical model (enclosing them) can be of some help towards a future and, possibly just partial, undoing what is known as the Gordian knot of artificial intelligence, that is, the symbolic mastering of commonsense reasoning in view of its mechanizing through computer algorithms and heuristic methods.

What follows is presented with the aim of trying to describe symbolically, without presuming too strong algebraic structures, what seems to surround the naïve expression, "*Questioning, Guessing, Telling, and Computing*," a scant but not totally wrong résumé for the concept of rationality, the old *Logos*.

This book could even be considered a trial towards a theoretical, but naïve, approach to rationality that, provided someday were transformed in a new experimental science, a physics of language and reasoning intermingling controlled experimentation and theory, with each reinforcing the other, could mean a paramount success towards a deepest knowledge of *Logos*.

It should still be added that this is not a book on certainties, but on questions and doubts. It is not full of technicalities and theorems, but consists in some reflections that, often supported by mathematical developments can, eventually, lead to new views of how to study ordinary reasoning and also fuzzy logic. In general, to know "why" something is what it is, it is better benefiting from previous analysis on "how" it can be managed; our "something" is ordinary reasoning, and our "analysis" concerns how it can be represented in a naïve symbolic form able to reveal some of its characteristics.

No final doctrine is tried to be imparted, and the core of what follows can be seen rooted in the conceptual troubles that can come from always presuming a strict mathematical framework, as is remarked in what follows. The agendas of this book and of mathematical logic, either crisp or fuzzy, are different; actually, at most they facilitate only a "view from the ground" of ordinary reasoning, and of classical and fuzzy logics. What is mainly offered is questioning of ordinary reasoning.

Concerning fuzzy logic, more than 50 years after Lotfi A. Zadeh introduced fuzzy sets, and also more than 20 after he introduced his first and brilliant ideas on *Computing with Words*, the moment seems to have arrived of theoretically rethinking fuzzy sets from their very grounds, plain language and ordinary reasoning, in short, looking at it from down to top, but not from top to down; that is, with a true scientific seriousness, and not only as either a kind of abstract playing, or as a complex of recipes just devoted to immediate applications, even if both existing approaches should not be disdained in the measure that they could help to see, down to top, fuzzy logic and computing with words.

Part I
Sowing Ideas

Chapter 1
Introduction

It can be said that reasoning introduces some kind of organization in thinking for directing it to a goal. What is required for a theoretical study of ordinary reasoning is to establish as simple and natural a framework as possible, in which symbols represent what corresponds to the setting into which the subject under study is inscribed. Without "representation" nothing of a scientific type seems to be possible. Science deserves representation and measuring, and from the very beginning of the scientific revolution (and even before), symbolic representation seemed to be a good instrument for acquiring scientific knowledge through measuring.

But symbolic representation and measuring need a mathematical framework, and not all mathematical structure is adequate to a given subject; nobody will try to study thermodynamics in, for instance, only the formal framework of a finite geometry, without mathematical analysis or experimentation.

1.1. Hence finding a suitable, or natural, framework for a subject is of paramount importance for, at least, not adding laws that cannot be immediately presumed for it. The relatively natural character of a particular mathematical framework depends on the kind of problems that are posed, and whose solutions are not only analyzed but processed. For instance, the framework of a Boolean algebra is a natural one for the reasoning dealing with precise statements, in which it is supposed that all the needed information on them is, in principle, available, and all the laws of such algebras are valid because each precise predicative word specifies a subset (is represented by it) in the corresponding universe of discourse. Inasmuch as its complement represents the negation of the word, each precise predicative word produces a perfect classification of the universe of discourse, and composed precise statements also produce partitions of such a universe. Boolean algebra is the undisputed natural framework of representation for computing with precise words; it facilitates a suitable calculus for mimicking reasoning with them.

Notwithstanding, and because neither the distributive laws, nor the law of perfect repartition, nor the equivalence between contradiction and incompatibility, are valid for reasoning on quantum physics phenomena, their statements cannot be

E. Trillas, *On the Logos: A Naïve View on Ordinary Reasoning and Fuzzy Logic*,
Studies in Fuzziness and Soft Computing 354, DOI 10.1007/978-3-319-56053-3_1

represented by sets but by functions in a Hilbert space whose structure is weaker than that of sets. Hence Boolean algebra ceases to be the natural framework for such a kind of reasoning, and weaker algebraic structures, such as orthomodular lattices, were taken for it instead of Boolean algebras. Analogously, Boolean algebras are not a natural framework for the analysis of the reasoning involving both precise and imprecise statements, in which the last cannot be specified by sets but by the membership functions of fuzzy sets. With them, for instance, the negation is not a "Boolean complement", and neither the principle of contradiction, nor that of the excluded-middle, can always be supposed to hold as they do with sets. No Boolean algebra and no orthomodular lattice can be taken as a natural framework for it even if, in some special cases, the weaker framework of a De Morgan algebra can be suitable.

Hence, for the analysis of ordinary reasoning that comprises all particular types of reasoning, a natural mathematical framework necessarily should be extremely weak and, in particular, weaker than Boolean and De Morgan algebras, than orthomodular lattices, and than the weakest standard algebras of fuzzy sets.

It should be pointed out that in the methods for "logically proving", such as the first one established in 1934 by Gerhard Gentzen, and called Gentzen's natural system, Boolean laws are not explicitly formulated, but implicitly used; these systems are for ruled deduction, that is, for formal deduction without jumps and whose conclusions are but "logical consequences". That is, its set (following from that of the given premises) satisfies the properties of a compact Tarski operator of consequences characterizing formal deduction. In ordinary reasoning it cannot be presumed that statements perfectly classify the universe of discourse; things are more complex, imprecision is pervasive, and degrees should be considered. Nevertheless, and for all this, arriving at a "computing" with both precise and imprecise words requires a calculus rooted in a very general and weak mathematical framework, able to be compacted in a stronger mathematical structure in some particular cases, for instance, to a Boolean algebra when the subject is computing with precise words, or to a De Morgan algebra in some particular cases with imprecise words.

The mentioned Boolean algebra, De Morgan algebra, orthomodular lattice, and standard fuzzy algebra frameworks are but models for the corresponding types of reasoning, and, as such, are just simplifications of the actual reasoning by trying to take into account those of its characteristics considered to be the basic ones. Of the four models, only the first can be considered definitive, because it reflects formal reasoning with precise words well; this formal reasoning is exactly what is done when doing such kinds of proving reasoning. For a reasoning that can be conducted with pencil and paper, to speak metaphorically, there is nothing else; the model is the reasoning and hence the frame is a fully natural one. Nevertheless, in the cases of De Morgan, orthomodular, and standard fuzzy algebras, the model is still provisory in the sense that there is no general agreement on its full suitability for the corresponding subjects and, for instance, some logicians prefer to use weaker structures than orthomodular lattices for the analysis of reasoning on quantum physics. Note that one thing proves that "this" follows from "that" (the so-called context of proof), but that a very different one is to find an unknown "that" to be proven after knowing the "this" (context of search). In any case, the "that" and the "this" are, respectively, the reasoning's conclusion and premise.

A model can be seen as the mockup an architect makes to visualize in three dimensions how the true building finally will be, or as a map that does not contain all that is in the mapped land, yet allows finding its places. But, neither the mockup nor the map is exactly the building or the land; they are just simplified representations of them. A basic theoretical task consists in finding a model allowing us to, at least, present the essential characteristics of what is being modeled, and allowing us to foresee more things; a model should arise from some knowledge of the corresponding subject, and a good one also allows finding novelties.

In the case of ordinary reasoning, models should arise from the currently scarce knowledge of it. Which is such knowledge, and how can it be improved and extended? This question could be seen similarly to how visual knowledge of the heavens in Copernicus' time passed to that after the telescope and Galileo, to Newton–Leibniz's invention of differential calculus and Newton's mathematical theory of gravitation, and, perhaps finally, to Einstein's mathematical model of general relativity, which today nobody can predict whether it actually models the full macrophysical reality, even if experimentation seems continuously to confirm its validity.

Each time, frontiers for new theories and for new experiments (usually also helped by new technology) either confirming or falsifying the former knowledge with their measurements were open. This is the form in which science typically advances through a continuous interlinking of mathematical models, technology, and experimentation, but without stating that things can be, definitively and exactly, identified with the model. Models are engines driving the only safe type of knowledge of the reality, the always uncertain scientific one; the rest is but metaphysics just based on pure abstract thinking in which the meaning of the employed words is taken as something universal and without considering their possible measurability. It is a kind of elemental reasoning coming from the old times where primitive metaphors were taken as reality, but not as ways for just reflecting on them, and that should be embodied and clarified in a theory of ordinary reasoning; it is a type of reasoning by analogy that, fueled by ordinary deduction, is able to create the brain's images and is very useful for producing emotion, exciting sentiments, and provoking ordered speculations. If it does not properly serve to describe reality deeply, it has, nevertheless, the power of directing the intellect towards creativity, the "last mystery" in the words of the writer, Stephan Zweig. Many of our current concepts have their roots in metaphor, and the metaphysic mode of reasoning is not at all contemptible.

1.2. To construct a theory on the meaning words show in plain language, let's say their linguistic meaning, it should be taken into account that meaning is not universally associated with a word in itself, but that words are usually context-dependent and purpose-driven. For instance, in the context of the positive integer numbers, "prime" has a precise meaning given by an "if and only if" definition, with a purpose restricted to the field of arithmetic, and related to expressing all natural numbers by the product of their prime divisors; but in a context of people the meaning of "odd" cannot be defined in such a precise form, it can only be described as a synonym of rare-person and perhaps used with some denigrating, or hilarious, purpose. It is not with the same meaning that it is said, "Seven is odd", that "He is odd". Nevertheless, in the setting of the positive integer numbers, "odd" has a precise meaning given by an if and only if definition, and, without a previous definition of what a "very odd number" can consist in, it is not possible to state and understand, "This number is very odd", although it is not necessary to add anything to the use of odd with either people or houses, and so on, to recognize immediately what the statement, "Such person/house/... is a very odd one", describes in a given context.

A look at a dictionary shows that it contains words mainly belonging to three categories: those whose use is imprecise in a given context and precise in another, those whose use is always precise, and those whose use is always imprecise, the second category of which contains a relatively small number of words. The meaning of words is always contextual or situational; precise words abruptly change their meaning when the modifier "very" affects them, and the imprecise ones are those that, once their meaning is captured and, if affected by "very", the meaning of the new expression is immediately captured. Without taking into account the purpose for its use, precise words have a rigid meaning, but that of the imprecise is flexible and how they are recognized after being affected by the linguistic modifier "very" seems to indicate that, at least for imprecise words, there are some qualitative variations of their use, and going from less to more, that the meaning of imprecise words is not static but flows in the universe of discourse.

As Ludwig Wittgenstein wrote, "The meaning of a word is its use in language", and, for instance, "odd" is used in the language of arithmetic by using it according to the rule, "n is odd" $\Leftrightarrow$ "the rest of dividing n by 2 is 1", but "big" is used in the interval [0, 10] according to the rules, (1) x is "less big than" $y \Leftrightarrow x \leq y$; (2) 10 is maximal for the relation "less big than"; and (3) 0 is minimal for such a relation. These rules allow several interpretations, such as the rigid one "x is big" $y \Leftrightarrow 7 \leq x$, and its flexibility is additionally shown by the fact that if x can be qualified as big, there is $\varepsilon > 0$ such that all numbers between x and $x + \varepsilon$ can also be qualified as big; if "7 is big", then also "7 + 0.0001 is big"; with it, any rigid use of "big" should be avoided. Often, the use of words can be described by instances of their application, or by rules (not always precise ones) describing how they can be used. Note that 5 is an odd number, but that adding unity to it (the smaller positive integer) what is obtained, $5 + 1 = 6$, is not an odd number; odd is not flexible among the positive integers.

Without capturing the linguistic relationship "less than", or its inverse "more than", it is unknown how the word's application flows along the universe of discourse, how its application to the elements in the universe varies, or how it is imprecisely used. Without knowing the prototypes (or maximal elements) the universe contains, and the antiprototypes (or minimal elements), provided they were to exist, it seems difficult to have instances of the total verification, or total nonverification, of the property to the elements the word names, and there is a lack of information on the word's use. Maximal refers to the nonexistence of elements showing more what the word carries, and minimal to the nonexistence of elements showing it less; without capturing the variations from less to more, prototypes, and antiprototypes, the meaning of a word is not well captured. For instance, understanding the use of "big" in [0, 10] as given by the set [7, 10], those elements in [0, 7) are antiprototypes of big, and those in [7, 10] are prototypes; the word's use is precise, and there are no more than prototypes and antiprototypes of it. Nevertheless, as shown, the word *big* has many imprecise uses in [0, 10].

Meaning is always related to a given universe of discourse X; hence to study the linguistic meaning of a word P, it is necessary to consider both the pair (X, P), and that P names a "property" that is recognizable, at least empirically, for the elements x of X; P is a predicative word in X. The word P says something about the elements in X that is translated by the elemental statements "x is P", resuming the information carried by P on X, with which the meaning of P flows along X, and eventually showing some vortices of concentration (the prototypes) and some of dissolution (the antiprototypes).

Once this can be formally established, the pair constituted by X and the relationship "less P than", allows us to define what a measure for the meaning is, as shown in what follows. When the use of P in X is precise, or rigid, there are only prototypes and antiprototypes, and the relationship "less P than" collapses with that "equally P than".

References

1. E. Trillas, I. García-Honrado, ¿Hacia un replanteamiento del cálculo proposicional clásico? Agora **32**(1), 7–25 (2013)
2. L. Wittgenstein, *Philosophical Investigations* (Basil Blackwell, Oxford, 1958)
3. M.E. Tabacchi, S. Termini, Fuzziness as an Experimental Science, in *A Passion for Multiple-Valued Logic and Soft Computing*, ed. by R. Seising, H. Allende (Springer, Berlin, 2016), pp. 41–53
4. S. Zweig, *The World of Yesterday: Memories of a European* (Viking Press, New York, 1943)
5. M. Black, *The Labyrinth of Language* (Pelikan Books, London, 1968)
6. S. Watanabe, *Knowing and Guessing* (John Wiley & Sons, New York, 1969)
7. D. Papo, Functional significance of complex fluctuations in brain activity: From resting state to cognitive neuroscience. Front. Syst. Neurosci. **8**, 112 (2014)
8. A. Tarski, in *Logique, Sémantique, Metamathématique*, vols. I, II. (Armand Colin, Paris 1972, 1974)

Chapter 2
Meaning as a Quantity

Considering the meaning of words is necessary for understanding both our own reasoning and that of others; with meaningless words people's plain language and ordinary reasoning would be for nothing, and communication impossible. Hence, a previous symbolic analysis of the meaning of words is important for better comprehending what ordinary reasoning is. In addition, meaning is a twofold concept: it has two sides, the qualitative and the quantitative, reflecting the situational use of words, namely that its use is context dependent and purpose driven. Often, and from a scientific point of view, only considering meaning from its qualitative side is insufficient; it is the quantitative that facilitates the degrees up to which words actually mean something in a given context. Meaning should be somehow measured.

2.1. Given a universe X in which a predicative word P is acting through the elemental statements "x is P", let's symbolically denote by $<_P$ the previously captured relationship "less P than",

$$x <_P y \Leftrightarrow x \text{ is less } P \text{ than } y,$$

translating the either empirically, or theoretically, recognized fact that x verifies less than y the property p named P. Then, the graph $(X, <_P)$ represents how P semantically organizes the universe of discourse X, and denotes a primary *qualitative meaning* of P in X. The inverse relationship "more P than" x is more P than y, coincides with $y <_P x$, and when it simultaneously holds $x <_P y$ and $y <_P x$, is that x is "equally P than" y, symbolically written $x =_P y$.

Note that, at the end, people's (intelligent) talking and telling tries to introduce some organization, or ordering, between the concepts/words under consideration, and for trying to answer the questions leading to telling something; in this respect, it seems natural to consider that the symbolic relation $<_P$ producing the qualitative

E. Trillas, *On the Logos: A Naïve View on Ordinary Reasoning and Fuzzy Logic*,
Studies in Fuzziness and Soft Computing 354, DOI 10.1007/978-3-319-56053-3_2

meaning is the way in which P semantically acts on X, introducing in it some organization. Usually, words are but names of concepts mastered though the meaning of the words.

It should be pointed out that as the relation $<_P$ is, when P is a word managed in plain language, empirically recognized, it implies some subjectivism that is, nevertheless, shared with others; $<_P$ is here presented as just a kind of primitive idea, as if it were "point" and "line" in the old Euclidean *Elements*.

It should be pointed out that the relation $<_P$ is not, in general, a linear or total relation; that is, there can exist pairs of elements x and y such that it is neither $x <_P y$, nor $y <_P x$; in this case, x and y are not meaning-comparable and are denoted by $x \, NC_P \, y$. For instance, in the toy example of the word P = big in $X = [0, 10]$, it is $x <_{big} y$ if and only if $x \leq y$, in the linear order $\leq$ of the real line; that is, $<_P = \leq$ is a total relation under which all the elements of the universe [0, 10] are big-comparable; hence x is "more" big than y whenever is $y \leq x$, and x is "equally big than" y whenever both numbers coincide, $x = y$. The qualitative meaning of "big" in [0, 10] is given by the graph ([0, 10], $\leq$), in which there is the unique maximal 10 and the unique minimal 0; obviously, in the real interval [0, 10], 10 is always a prototype of big, and 0 is always an antiprototype. In the same interval, calling "medium" the property of being around 5, the situation is different provided, for instance, the prototypes were those x in [4.9, 5.1], the antiprototypes were those x in [0, 3] U [7, 10], and

$$x <_{medium} y \Leftrightarrow x \leq y \leq 4.9, \text{ or } 5.1 \leq y \leq x,$$

in which case there are elements, such as 4 and 6, that would not be comparable, but isolated. Note that provided the prototypes were the elements in the open interval (3, 7), and the antiprototypes those in [0, 3] U [7, 10]; then the use of "medium" would be precise, and given by the necessary and sufficient condition (definition),

$$\text{'}x \text{ is medium' if and only if } 3 < x < 7.$$

Of the symbolic relation $<_P$, it can be easily accepted that it is always reflexive, such that $x <_P x$ for all x in X, but not, for instance, that it is a partial order; symmetry, antisymmetry, transitivity, and so on cannot always be supposed as properties $<_P$ holds.

Once a graph $(X, <_P)$ is recognized as a qualitative meaning of P in X, the mappings $m_P: X \to [0, 1]$, verifying the axioms:

(1) $x <_P y \Rightarrow m_P(x) \leq m_P(y)$
(2) x is maximal in the graph $\Rightarrow m_P(x) = 1$
(3) y is minimal in the graph $\Rightarrow m_P(y) = 0$

can be defined, and called measures of the meaning of P. Note that the interval of values [0, 1] can be changed by any closed interval of the real line, and taking its extremes instead of, respectively, 0 and 1. As in the case of probabilities, these three axioms do not allow us to specify a unique measure, but more information on the measure's contextual characteristics is necessary for it.

For instance, in the former case of "big" with qualitative meaning given by $([0, 10], \leq)$, provided it were known that the measure should be linear, $m_{\text{big}}(x) = ax + b$, from (2) and (3), it follows the single measure $m_{\text{big}}(x) = x/10$, also verifying (1) because it is a nondecreasing function; but provided it were contextually known that the measure should be quadratic, $m_{\text{big}}(x) = ax^2 + bx + c$, because then it should be $c = 0$, $100a + 10b = 1$, and $2ax + b \geq 0$ for all x in [0, 10], many quadratic measures would be possible, for instance, $x^2/100$, $x^2 + 0.99x$, $-2x^2 + 20.1x$, and the like.

In any case, a full use of P in X associated with the relationship "less P than," is given by the quantities $(X, <_P, m_P)$; each time one of these quantities is specified, a full meaning is scientifically designated a quantity, or specified, for P in X. It should be noted that all measures preserve the relationship "equally P than":

$$x =_P y \Leftrightarrow x <_P y \text{ and } y <_P x \Rightarrow m_P(x) \leq m_P(y) \text{ and } m_P(y) \leq m_P(x), \text{ or } m_P(x) = m_P(y).$$

In this form, each meaning P can have in X can be seen as a quantity.

What about precise words? Because a precise word P specifies, by definition, a subset **P** in the universe of discourse, all elements in **P** are prototypes and it is $x =_P y$, for each pair x, y of them, and those in $\mathbf{P}^c$ are antiprototypes with each pair also being equally not P; hence, the first are maximal and the second minimal. The universe is partitioned in the maximal and the minimal, each class having equal measure for all its elements. Consequently, there is just a single measure specified by

$$m_P(x) = 1 \text{ if } x \text{ is in } \mathbf{P}, \text{ and } m_P(x) = 0 \text{ if } x \text{ is in } \mathbf{P}^c,$$

which is just the characteristic function of **P**. The reciprocal is obvious, and then $m_P^{-1}(1) = \mathbf{P}$; thus, precise words P are specified by the single graph $(X, =_P, m_P)$, with the measure only taking its values in the subset {0, 1} of the interval [0, 1].

Note that it can be cases of words without prototypes or antiprototypes, also partitioning the universe in several subsets of equally P elements, but with measures that, constant in each of these parts, have values not 0 or 1, but only values in the open interval (0, 1); of these words it could be said that their use, or meaning, is pseudo-precise in X. Changing the measure m by $m/\text{Sup } m$, provided $\text{Sup } m = \text{Max } m < 1$, at least an element with measure one would appear.

2.2. Once a measure m_P is specified, it defines in X a new, and linear, relation $\leq_m$,

$$x \leq_m y \Leftrightarrow m_P(x) \leq m_P(y),$$

that is larger than the relation $<_P$, because

$$x <_P y \Rightarrow m_P(x) \leq m_P(y) \Leftrightarrow x \leq_m y.$$

That is, $<_P \subseteq \leq_m$, and, for coinciding, $<_P$ should be linear; nevertheless, in general, both relations are not coincidental, and the second being larger than the first gives a "larger and linear new meaning" $(X, \leq_m, m_P)$ that, reached after m_P is known, can be called a *working meaning* of P in X; noncomparable elements under $<_P$ are always comparable under $\leq_m$.

It can be said that measuring enlarges meaning by linearizing the qualitative meaning $<_P$ up to $\leq_m$ Once a working meaning substitutes the qualitative meaning, new working prototypes can appear, and working antiprototypes, those x in X, respectively verifying $m_P(x) = 1$ and $m_P(x) = 0$, that can be more than the original prototypes and antiprototypes. This deserves some comments.

The first comment refers to how a fuzzy set and its several membership functions can be understood. Up to now it is not well known what a fuzzy set in X with linguistic label P actually is, and it is usually confused with just one of their membership functions; but because, unless the linguistic label's use is precise in the universe, there is not a single membership function characterizing the fuzzy set, such identification is actually a one-to-many correspondence. It is usual to consider that fuzzy sets are represented by the functions in $[0, 1]^X$, something only useful for purely mathematical purposes and once a membership function is specified or, better, designed. It is not the same with sets that can be quietly identified with the functions in $\{0, 1\}^X$ thanks to the unicity of their measure, or characteristic function. The concept of a fuzzy set is not only a matter of degree, as its originator Lotfi A. Zadeh likes to say, but its membership functions are a matter of careful design. To capture what is actually a fuzzy set is a question that should be initially posed in the setting of plain language.

In the first place, such a question refers to the empirical fact that predicative words "collectivize" in the universe of discourse, that they generate "linguistic collectives" well anchored in plain language. For instance, in the universe of London's inhabitants, the word "young" generates the linguistic collective of "young Londoners"; in the universe of the real numbers the word "big" creates the linguistic collective of "big numbers"; in a universe of buildings the word "high" creates the linguistic collective of "high buildings", and so on. Obviously, linguistic collectives are well understood by the speakers, but are kinds of gaseous or cloudy entities for which no criteria of individuation are known, and, inasmuch as, following W.V.O. Quine, "There is no entity without identity", linguistic collectives should be approached through ways of which, right now, the one only at hand comes from the quantities specifying the meaning of the word at each universe of discourse.

It could be said that the linguistic collective P generates in X is in the qualitative state $(X, <_P)$, and that to each qualitative state it corresponds to being in several quantitative states each given by a measure m_P. Each full meaning given by a quantity $(X, <_P, m_P)$ shows both a qualitative and a quantitative state of the linguistic collective; it is a state of the collective. States reflect the available information on the qualitative and the quantitative use of P in X.

It is a view reflecting the several situations in which the collective can be seen, and under it, a fuzzy set in X is nothing else than the linguistic collective generated by its linguistic label. It just consists in renaming the collective of P in X as the fuzzy set labeled P. In the case where the linguistic label P is precise in X, the collective "vitrifies" in the set specified by P in X; the collective has just the qualitative state $(X, =_P)$, and just the quantitative state given by its characteristic function m_P giving the vitrified, or crisp, quantitative state $m_P^{-1}(1)$. Precisely used linguistic labels are in a single state.

When the linguistic label is imprecisely used in X, the collective or fuzzy set will have as many qualitative states as qualitative meanings $(X, <_P)$ can be recognized, and each measure for them is nothing else than a membership function for the fuzzy set labeled P. Note, for instance, that there is no difference between the former measures m_{big} and the membership functions that can be attributed to the fuzzy set in [0, 10] labeled "big".

Conceptualizing a fuzzy set as just an abstract entity, or concept, existing in language, helps to refine Zadeh's intuitive view that each membership function is an extensional meaning of the corresponding linguistic label.

In the imprecise case, the designer of a membership function of a fuzzy set in X with linguistic label P should proceed by taking into account all the information available to her or him on the use of P in X that is, usually, incomplete, sometimes not containing all the relations $<_P$, and to which often the designer still adds some reasonable hypotheses on the shape of the membership function that can be suitable for the current problem. For instance, and in the toy example of the fuzzy set in [0, 10] with linguistic label "big", the designer could consider that, with her or his current scarce information, the best that can be done is to take the simple and above-mentioned linear measure $x/10$, or if the designer can suppose it should be quadratic, just take its square $(x/10)^2 = x^2/100$. In short, the designer is often limited to consider some (possibly scarce) information of how P behaves in X, and some characteristics of the current problem for which the design is done. For instance, by estimating if the measure of "5 is big" should be 0.5, clearly less than 0.5, or clearly bigger than 0.5, and so on.

Hence in the praxis of fuzzy logic it cannot always be supposed that a membership function is actually a measure but, in the best case, that the membership function is a universal approximation of some measure; that is, a designed membership function μ_P could be seen as a good enough one provided some measure m_P exists such that, for instance, it would verify

$$|m_P(x) - \mu_P(x)| \leq \varepsilon, \text{ for all } \varepsilon > 0, \text{ and all } x,$$

or, if it is possible, that minimizes the function

$$\text{Sup } |\text{m}_p(x) - \mu_P(x)|.$$

Anyway, and in each case, this requires previously counting with a measure m_P that makes the membership function μ_P unnecessary. Provided it were proven that such a criterion of approximation actually exists, perhaps a nice existential theorem for characterizing good membership functions could be obtained and become useful for the processes of designing them.

In short, and in general, the designed membership functions only can be seen as potential approximations to measures, with which only a working meaning $\leq_\mu$ is often available and, being a linear relation (a partial order indeed), cannot always coincide with the qualitative meaning. In praxis, the modeling by fuzzy sets is just an uncertain approximation to the meaning of their linguistic labels, and true measures are but "ideal" membership functions such as the uniform probability 1/6 reflects an "ideal die" in probability theory; for this reason, membership functions should be carefully designed on the basis of the best information available on the behavior of P in X, and with the best possible reasonable hypotheses on their shape coherently with the requisites of the current problem.

It should be pointed out that, in plain language and ordinary reasoning, it is sometimes difficult and even unnecessary, to attribute numbers for measuring the meaning; for instance, there are no rare expressions such as "It is highly possible that John is rich," "The degree up to which Jane is wise is up to the middle," or "Susan is extremely intelligent." Hence, in praxis it could be suitable to substitute the interval [0, 1] in which the measure's values range by sets such as the subintervals in [0, 1] or the fuzzy numbers in [0, 1] or even a set of linguistic labels. With them, the meaning of the former examples can be measured by, respectively, a fuzzy number μ_{high}, the interval (0.7, 1], and the word "extremely". Because fuzzy numbers and intervals are not totally ordered, these kinds of values can still present the advantage, when $<_P$ is not linear, of having more possibilities for the coincidence between the qualitative and the working meanings. Anyway, the set of such values should be endowed with some algebraic ordering, necessary for defining a measure, and contextually coming from their uses.

In science complex numbers are sometimes taken to measure some variables and, in the same vein, instead of the unit real interval [0, 1], the complex unit interval $\{a + ib; a, b \text{ in } [0, 1]\}$ can be taken by submitting the measure to verify the three former axioms but with its values in the complex unit interval, endowed with the natural (partial) order of complex numbers:

$$a + bi \leq^* c + di \Leftrightarrow a \leq c \,\&\, b \leq d,$$

with maximum and minimum, respectively, $1 + i$, and $0 + 0i = 0$, giving a nonlinear working relation $\leq_m = \leq^*$ for all measures m; now the working meaning is not a total or linear relation, but a partial ordering. Because the complex unit interval can be seen as isomorphic with the set of closed subintervals of [0, 1], taking one or another as the range for the measures does not matter. Once accepted as a natural way of ordering such complex numbers, intervals, fuzzy numbers, words, and the like, as well as which are their respective maximum and minimum, it is necessary to take the suitability for the considered problem of the calculus with them into account, something that should be done at each concrete problem for having, provided it were the case, a computation with the corresponding words, or statements.

2.3. Let's apply what has been said to the (debatable) concept of truth, deriving from the word "true" applied to statements; it is usual to identify expressions such as "What you say is true" and "You are saying the truth," but here "true" is understood as a predicative word, and not as naming the concept "truth" which can be seen as its mother-predicate. Concepts are but abstractions generated after using a word as its mother-predicate to either physical or virtual objects, and usually once such a word migrated between several universes and suffered some more or less slight modifications in its respective meanings. It is not "tall" that comes from the concept of "tallness", but this (abstract) concept was generated after applying "tall" to several collections of objects including trees, mountains, people, and so on, and passing from one to another by analogy; each time, "tall" refers to some particular objects, but "tallness" refers to all of them. In the same vein, "true" refers to each statement, but "truth" refers to all of them.

Truth is a concept that, understood as an absolute and universal one, has been conducive to aberrations and, indeed in both its past and present human history, carries terrifying consequences coming from such an understanding of "Truth" with a capital letter.

It should be pointed out that the qualitative meaning of a word applied to the limiting case of a universe of discourse just consisting in a singleton $\{x\}$, is reduced by the reflexive property to the minimum relation $<_P = \{(x, x)\}$, of which nothing can follow, and, first of all, it suggests some comments referring to when it can be said that the meaning of a word is "metaphysical", that a word is meaningless in some universe of discourse and hence cannot generate a concept; that, out of the singleton, it does not collectivize.

Once a qualitative meaning $(X, <_P)$ is captured, it can be said that P is pseudo-measurable in X, because it is what allows defining measures for P and there exists at least the one defined by assigning 1 to the prototypes, 0 to the antiprototypes (provided both exist), and a fixed and common value to the other elements in X, for instance, 0.5, even if these measures could have nothing to do with the context in which the word is used and consequently do not reflect anything but a lot of ignorance on P in X. When at least a measure related to the context can be specified, it can be said that P is measurable in X, or, perhaps better, is effectively measurable; for instance, "big" is effectively measurable in [0, 10]. When no relation $<_P$ can be captured, it can be said that the word is meaningless in X, that its use is currently metaphysical because it is not even pseudo-measurable, and measurability is a main characteristic science requires for the predicates it manages. It is worth to remember Lord Kelvin's shortened statement, "If you cannot measure it, it is not Science".

That P is meaningless in X does not imply that it should also be meaningless in all possible universes of discourse; for instance, "big" is measurable with real numbers, but meaningless if applied to dreams, because it seems actually impossible to recognize previously when a dream is "less big than" another one. Nevertheless, what cannot be inferred is that a nonmeasurable "big" can never be useful for reaching some new idea; stating "a dream is big" could serve as a useful metaphor, or analogy, able to induce the study of something related to dreams; P is meaningless in X does not imply that P cannot excite someone towards a further searching, but psychological matters are beyond what we are trying to analyze here, even if it is manifest that metaphorical thinking is important for conducting creative reasoning; provided, of course, it were to be produced jointly with having some knowledge of the corresponding subject as pointed out before with the Kekulé example. It should be noted that, although for scientific purposes measurability is essential, neither all that is relevant for ordinary reasoning is measurable, nor what is measurable always important for ordinary reasoning.

What about the meaning of the word T = true in a set X [P] whose elements are the elemental statements x is P, for x in X, and after knowing that P is measurable in the universe X? The first to be captured is the relationship "less true than":

$$x \text{ is } P \text{ is less true than } y \text{ is } P \Leftrightarrow x \text{ is } P <_T y \text{ is } P,$$

and the second is recognizing which are the maximal and minimal statements in the graph (X [P], $<_T$), provided they were to exist, that is, specifying a qualitative primary meaning of T in X [P]. Note that, instead of X and P, it could be considered, for instance, the union universe X U Y, and the two words $\{P, Q\}$, to capture

$$x \text{ is } P \text{ is less true than } y \text{ is } Q,$$

for x in X and y in Y, with P acting in X and Q in Y with respective qualitative meanings (X, $<_P$), and (Y, $<_Q$); but, for simplicity, and even if several words could be taken into account, only one universe and a single word are considered right now.

True = T names a property of the elements in X [P] referring to the actual verification of the property named P for the elements in X, the reality of the statement "x is P". That is, the character of "true", a statement "x is P" can show, is directly related to the verification by x of the property named by P (as more P is x, more true is x is P), and that, hence, some relation between the qualitative meanings $<_P \subseteq X \times X$, and $<_T \subseteq$ X [P] $\times$ X [P], should exist. It is supposed for it that "If $x <_P y$, then x is $P <_T y$ is P". It seems, consequently, that for specifying a measure of T it should be linked with one of P, and the question is how it can be done.

Inasmuch as it is m_P: $X \to [0, 1]$ and m_T: $X [P] \to [0, 1]$, let it be t: $[0, 1] \to [0, 1]$ a nondecreasing mapping such that $t(0) = 0$, and $t(1) = 1$, namely an order morphism of the ordered unit interval ([0, 1], $\leq$), under which conditions can it be m_T (x is P) = (t o m_P) (x) a measure of T?

Because $x <_P y$ implies x is $P <_T y$ is P, it is

$$m_P(x) \leq m_P(y), \text{ and } t(m_P(x)) \leq t(m_P(y)), \text{ or } m_T(x \text{ is } P) \leq m_T(y \text{ is } P),$$

that is, the first axiom of a measure for T is verified. For what concerns the other two axioms relative to maximal and minimal elements, it is immediate that if x is maximal (resp., minimal) for $<_P$ it follows m_T (x is P) = $t(1) = 1$ (resp., m_T (y is P) = $t(0) = 0$), but the question seems to be opened when "x is P" is maximal, or "y is P" is minimal for $<_T$ without necessarily implying that x or y are, respectively, maximal and minimal for $<_P$. Nevertheless, by supposing that T is "coherent" with P, that is,

x is maximal (resp., minimal) for $<_P$ if and only if x is P is maximal (minimal) for $<_T$, the problem is solved. Because coherence is not a bizarre condition, under it, t o m_P is a measure of T.

It is classically and usually said that m_T (x is P) is a "degree of truth" of x is P, thus when it is $m_T = t$ o m_P, the degrees of truth are obtained through a "truth-function" t. Then it should be noticed that it is m_T (x is P) = $m_P(x)$ for all x in X, provided $t = \mathrm{id}_{[0, 1]}$, a case in which the degree of truth of "x is P" just coincides with that in which x is P, as it is classically understood. Were, for instance, $t(x) = x^2$, it would be m_T (x is P) = $m_P(x)^2$, and then, if x is P with degree 0.7, "x is P is true" is with degree 0.49; if it were $t(x) = x^{1/2}$, then x is P would be true with degree $\sqrt{0.7}$ = 0. 836. Of course, the truth-function t should be chosen in each case according to the information available on the characteristics of the current situation, and by searching if, under them, the degrees of T should be lower or bigger than those of P, if t is, respectively, contractive ($t(x) \leq x$), or expansive ($x \leq t(x)$) for the x in X.

Summing up, provided T were coherent with P, truth-functions t would allow us to obtain measures of T from those of P, and whenever t is one to one and onto (bijective, an order automorphism of the unit interval) there are no more statements with measure one or zero for T than, respectively, the prototypes and the antiprototypes of P; nevertheless, a nonbijective truth-function, such as

$$t(x) = 0 \text{ for } x \text{ in } [0, 0.4],$$
$$t(x) = 1 \text{ for } x \text{ in } [0, 6, 1], \text{ and}$$
$$t(x) = 5x - 2 \text{ for } x \text{ in } [0.4, 0.6],$$

shows more statements with one or zero degrees of true than the prototypes and the antiprototypes of P.

2.4. Plain languages are not the creation of a single person, but are slowly generated by linguistic interactions between groups of people; plain languages are socially constructed over the course of time, and words acquire meaning along such interactions. Without sharing common meanings, people cannot actually understand each other, and communicating is very difficult if not impossible; linguistic meaning is a social construction that, to some extent, deserves a symbolical analysis to see how people can arrive at sharing a common meaning for words.

For such a goal, suppose that a person p_1 manages a word P in X under the qualitative meaning given by the graph $(X, <_P{}^1)$, and that this person utters to another p_2 elemental statements "x is P". Person p_2 will understand what p_1 is saying just provided she either captures the relation $<_P{}^1$, or, at least a nonempty part of it. That is, she manages P with a qualitative meaning $(X, <_P{}^2)$ such that the intersection of the respective relations $<_P{}^1$ and $<_P{}^2$ is not empty; on the contrary, if such an intersection is empty, p_2 cannot understand what p_1 tells her. To understand what p_1 says, p_2 can accept as the meaning of P, either all $<_P{}^1$, or a part of it, that will be denoted by $<_P{}^2$ and become a common meaning of P for both p_1 and p_2. Provided a third person p_3 enters later on in the conversation, the situation would be repeated, and a common meaning by the three people is reached by either the intersection of the three corresponding meanings, or by accepting a part of that meaning previously accepted by the first two, and so on.

Of course, when time passes, P can migrate from X to another universe Y, and so on; with it, and at the end, one or several meanings can result associated with the word P. Of course, it can happen that p_2 has no meaning for P, or that P is an unknown word to her, and, then, either the communication between p_1 and p_2 is impossible, or p_1 explains to p_2 the meaning of P by, for instance, practical exemplification, such as when exemplifying "the door is closed", by opening the door and saying "open", and closing the door and saying "closed" through a practical, or visual, description of the meaning of P.

Concerning how a quantity reflecting a full meaning of P in X can finally appear, a way for it could consist in aggregating the several measures assigned by each successive person in such a way that the aggregation preserves the verification of the three axioms a measure should satisfy. For instance, given the measures each of n people ($m_P{}^k$, $1 \leq k \leq n$) assigns to its respective qualitative meaning ($<_P{}^k$), the function

$$m_P(x) = \min\left(m_P{}^1(x), \ldots, m_P{}^n(x)\right),$$

is a measure for the not empty intersection $<_P$ of all relations $<_P^k$ because:

(1) $x <_P y \Leftrightarrow x <1_P y \; \& \ldots \& \; x <_P^n y \Rightarrow m_P^1(x) \leq m_P^1(y) \; \& \ldots \& \; m_P^n(x) \leq m_P^n(y)$ $\Rightarrow \min(m_P^1(x), \ldots, m_P^n(x)) \leq \min(m_P^1(y), \ldots, m_P^n(y)) \Leftrightarrow m_P(x) \leq m_P(y)$.
(2) If x is a maximal for $<_P$, then it should be a maximal for all the $<_P^k$. Hence, $m_P(x) = \min(1, \ldots, 1) = 1$.
(3) If x is minimal for $<_P$, then it should be minimal for at least one $<_P^k$, and because between the brackets of the min at least one zero will appear, it is $m_P(x) = 0$.

Of course, min is not the only possible function allowing that proof; any n-place function being nondecreasing at each place, taking the value one for the argument (1, …, 1), and the value zero for those arguments containing at least one zero, permits repeating the proof. In sum, there are many ways for reaching a common full meaning, and some of them can be symbolically represented and linguistically interpreted.

It should be pointed out that what has been developed does not pretend to be the only way in which all common meaning is reached; it simply tries to show that there are ways for it that are describable through a symbolic formalism. Anyway, the full theoretical problem still remains an open one in which the role analogy plays in it should be considered, and for whose solution the concourse of some evidence not only obtainable by means of clever observation seems necessary, but especially by well-designed controlled processes of experimentation on how meaning evolves in plain language. Possibly such types of problems require a new experimental methodology for studying language, supported by what computer technology can facilitate today for reaching it.

The importance words' migration can have was mentioned before, therefore let's add something in this respect; at the end, everything that has been reported in a symbolic form has at its back Ludwig Wittgenstein's ideas on meaning as use, family resemblances, and language games, presented in his (posthumous) second book, *Philosophical Investigations*.

In this respect, an example could be in order: the word "big" applied to numbers can be seen as a migration of "tall" applied to people after the centimeter to state numerical height is introduced. Indeed, if John's height is 190 cm, because 190 is a big number between those in [0, 200], interpreting "John is tall" as the linguistic evaluation of John's height as "big", is seen the migration between people and numbers of "tall" into "big". In an analogous way, "small" can be imagined as a migration of "short" from people to numbers, and further said that these pairs of words show a kind of linguistic family resemblance going from linguistically playing with people to playing with numbers. Perhaps this comment could help to find a (mathematical) way to analyze a linguistic phenomenon deeply under which people often learn how to understand and manage words, how to play with them in plain language.

Some naïve comments on syntax and semantics aspects are still in order, and once clearly said that if both aspects are basic for managing a plain language well, the meaning of statements not syntactically well constructed is often well understood, as happens, for instance, with children's speech. But, without capturing the meaning of statements, its semantics, there is no way of understanding them; for instance, sometimes children signalize something of what they refer to and thus what they are trying to say is recognized. For comprehending a language, semantics is essential, but syntax is often, although important, an accessory; for instance, in a talk on pigeons it does not matter very much if someone ignores how "pigeon" is written, but what is essential is not confusing a pigeon with an eagle, or a raven, or a vulture, and so on, and less again with something not being a flying bird, that is, recognizing the meaning of "pigeon" in the universe of "flying birds".

In logic, complex statements are supposed to be constituted of elemental statements "x is P", "y is Q", and the like, joined by connectives such as "and", "or", "not", and, sometimes in the conditional form "if/then", or affected by the only quantifiers "exists" and "all". But in the case of plain language, and in addition, those statements are often affected by linguistic modifiers such as "they are 'very'" and "more or less", and linguistic quantifiers, such as "many", or "several", or "few", as it is done with fuzzy sets. Then, and both in classical logic and in fuzzy logic, the meaning of a complex statement is supposed to be captured from the meaning of its components, after knowing which are the meanings of conjunction, disjunction, negation, conditionals, and so on, and presuming they are previously and contextually specified by either characteristic or membership functions. Nevertheless, in ordinary situations it is often the case of first capturing the meaning of a full complex statement, and second those of its components thanks to the contextual meaning the full statement facilitates for them.

Anyway, and for a theoretical study, the, let's say classical, way of constructively studying the meaning of complex statements through its parts cannot be avoided, hence, instead of a (currently unknown) synthetic and systematic procedure for directly analyzing the meaning of complex statements, it seems suitable, for a first symbolic study, to attempt one of analytic type in which the meaning of the components already takes into account the context in which they are inscribed. The next chapters are devoted to such a task, and, as always in this book, without presupposing laws that can reduce the analysis to be enclosed in a restricted, and up to some extent artificial, mathematical framework.

2.5. Finally, it should be pointed out why here it is preferable to say "ordinary, or plain language", instead of the usual expression "natural language", a preference just coming from the adjective "natural" used to show its opposite character to the axiomatic and "artificial languages" typical of formal logic and basically used to expose proofs clearly being sure that they contain no jumps. Because natural/plain language, during the process of representing, has to adapt itself to the pressures of its own capabilities and the corresponding representing goals, it never finishes being as "natural" as it had, presumably, begun; and this is because at the beginning it was still not committed to a specific work. There is no way of faithfully representing language and reasoning except with plain language; there can be thinking without language, but not human reasoning without language.

In addition, each plain language is permeated by traits coming from the cultural environment of their speakers; even English, today the almost universally common language in science, at least, is currently being influenced by the ways of speaking and writing it by people whose native language is not English, or whose native country is not an English-speaking one. In sum, plain language is not properly "natural" in the same sense that the brain or the Amazon forest is natural, but a result of many cultural, historical, geographical, and intellectual influences. Ordinary people not having attended school and not writing or reading their own plain language well, still speak it well enough and communicate easily with people educated in universities; the first perhaps don't manage it perfectly from the point of view of syntax, but well enough from that of semantics for conducting the second to capture the meaning of what they express.

It should be remarked that the same ownership title of Spanish belongs to a farmer in Mexico, or a Chilean living in Paris, the President of Argentina, the King of Spain, or a writer who won the Nobel Prize by his work originally written and published in Spanish. Plain languages are a shared common property of, at least, all its native speakers, and are among the most complex dynamical systems today science is faced with; new perspectives are needed for its scientific domestication.

In the end, mechanizing a plain language towards undoing the Gordian knot of artificial intelligence cannot be done without knowing its use, and once represented in a form preserving its flexibility. A form not previously constrained by logical laws not necessarily holding in plain language, as is the case with the commutative law of conjunction ($p \,\&\, q = q \,\&\, p$), almost always accepted in logic, but that, because time often intervenes in plain language, is a law that cannot be universally supposed, and as shown, for instance, by "She enters the room and starts crying", and 'She starts crying and enters the room", two statements depicting different situations, and whose identification can consequently lead, later on, to committing some mistakes.

In the same vein is the identification of conditional statements "If p, then q" ($p \to q$), with the affirmative statements "not p, or q" ($p' + q$), regardless of, for instance, whether the antecedent's negation (not p) can be suitably described and rightly represented after capturing its meaning in the corresponding setting, or if the underlying formal framework allows us to follow q from p and $p \to q$ (modus ponens, MP), as happens in the frame of Boolean algebras, but neither with p $q = p' + q$, in those of De Morgan algebras and ortholattices, nor generally in those of the standard algebras of fuzzy sets, where the inequality $p \cdot (p' + q) \leq q$, formally representing the rule of modus ponens, does not hold for all the pairs p, q, and all representations of "and" ($\cdot$), 'or' (+), "not" ($'$), and "it follows" ($\leq$).

For instance, if such inequality is considered in a setting endowed with a De Morgan algebra framework, by taking any element p, and $q = 0$, it follows that $p \cdot (p' + 0) \leq 0 \Leftrightarrow p \cdot p' = 0$; that is, p is one of the Boolean elements in the De Morgan algebra and, hence, the inequality does not hold for any pair p, q. Of course, in a setting endowed with a Boolean algebra's structure, and because both the distributive and the noncontradiction laws hold in it, it follows that $p \cdot p' + p \cdot q = 0 + p \cdot q = p \cdot q \leq q$, and the MP-inequality $p \cdot (p' + q) = p \cdot q \leq q$, holds for all pairs p, q.

In the case where the setting is of fuzzy sets μ, σ, and so on, endowed with the framework given by a standard algebra, the MP-inequality should be expressed by the functional inequality

$$\mu \cdot (\mu' + \sigma) \leq \sigma, \text{ or } T\, o\, (S\, o\, (N\, o\, \mu \times \sigma) \leq \sigma,$$

with T a continuous t-norm representing the linguistic "and", S a continuous t-conorm representing the linguistic "or", and N a strong negation function for the linguistic "not", whose study can be reduced to solving the (numerical) functional inequality

$$T(a, S(N(a), b)) \leq b, \text{ for all a, b in } [0, 1],$$

in which $b = 0$ ($N(0) = 1$) implies $T(a, N(a)) = 0$, showing the MP-inequality cannot hold for all pairs of numbers a, b, except for some triplets (T, S, N) such as the one constituted by Lukasiewicz's t-norm, $W(a, b) = \max(0, a + b - 1)$, its dual t-conorm, $W^*(a, b) = \min(1, a + b)$, and the strong negation function, $N = 1 - \text{id}$:

$$\begin{aligned} W(a, W^*(1-a, b) &= W(a, \min(1, 1-a+b)) \\ &= \max(0, a + \min(1, 1-a+b) - 1) = \min(a, b) \leq b, \end{aligned}$$

for all pairs of numbers a, b in [0, 1]. Notice that with $T = \min$, $S = \max$, and $N = 1 - \text{id}$, the inequality $\min(a, \max(1 - a, b)) \leq b$ does not hold for the pairs $a = 0.\ 5$ and $b = 0$; and, analogously, with the triplet given by $T = \text{prod}$, $S = \text{prod}^* = \text{sum} - \text{prod}$, $N = 1 - \text{id}$, the inequality is $\text{prod}(a, \text{prod}^*(1 - a, b) = a \cdot (1 - a + b - (1 - a) \cdot b) = a \cdot (1 - a + a. b) \leq b$, and it also does not hold for $a = 0.5$ and $b = 0$.

2.6. To end this chapter, and even if later on the subject is reconsidered, let's advance something of a preliminary character on the use in ordinary language of the words "uncertain", and "probable", that is, on describing and differentiating their respective qualitative and quantitative meanings; it is a subject that, linked with what has been presented on the meaning of words, is in touch with the debate between the two main interpretations of the probability mathematical concept, the objective and the subjective. The first comes from an observed convergence of the outcomes' frequencies in random experiments whose possible outcomes can be well described, and the second from the experienced opinion of a "rational person" assigning a priori probabilities to events, that, once transformed into a posteriori probabilities by means of the Bayes formula, helps whoever, based on them, wants to take the risk of betting some money on their appearance. In both cases, nevertheless, a common belief is shared on the actual possibility of perfectly classifying the universe by a union of disjoint classes, something necessary for posing the additive law of probability in which both interpretations coincide, and that, jointly with assigning a probability equal to one to the sure event, allows us to obtain the law for the probability of the negation of outcomes.

Notwithstanding, the understanding of events as subsets, also underlying both interpretations, corresponds to naming the events by precise words, something that is not always the case in ordinary reasoning expressed in plain language, as it is not always to assign numerical probabilities as can be exemplified by typical utterances such as, "It is highly probable than John is rich," or "It is improbable that Laura will join us," and so on. The meaning of "probable", the mother-predicate of the probability's concept, is still to be clarified in plain language.

Because the meaning of "uncertain", mother-predicate of the uncertainty's concept, is also not clarified, and because uncertainty, as with imprecision in language, permeates almost all branches of science, there is confusion between probability and uncertainty deserving to be clarified. All these questions are considered further.

References

1. E. Trillas, C. Moraga, S. Termini, A naïve way of looking at fuzzy sets. Fuzzy Sets Syst. **292/C**, 380–395 (2016)
2. E. Trillas, On a Model for the Meaning of Predicates, in *Views of Fuzzy Sets and Systems from Different Perspectives*, ed. by R. Seising (Springer, Berlin, 2009), pp. 175–205
3. E. Trillas, R. Seising, On meaning and measuring: A philosophical and historical view. Agora **34**(2), 1–23 (2015)
4. W.V.O. Quine, *Philosophy of Logic* (Prentice-Hall, New York, 1970)
5. J. Bairwise, J. Perry, *Situations and Attitudes* (The MIT Press, 1983)
6. L. Wittgenstein, *Philosophical Investigations* (Basil Plackwell, Oxford)
7. P. Halmos, *Naïve Set Theory* (Van Nostrand, New York, 1960)
8. E. Trillas, On the use of words and fuzzy sets. Inf. Sci. **176**(11), 1463–1487 (2006)
9. E. Trillas, L. Eciolaza, *Fuzzy Logic* (Springer, Berlin, 2015)
10. B. Russell, *The Problems of Philosophy* (Williams & Norgate, London, 1912)
11. E. Trillas, S. Guadarrama, Fuzzy representations need a careful design. Int. J. Gen Syst. **39**(3), 329–346 (2010)
12. E. Trillas, C. Moraga, Reasons for a careful design of fuzzy sets, in *Proceedings of the EUSFLAT* (2013), pp. 140–145
13. E. Trillas, I. García-Honrado, Unended Reflections on Family Resemblances and Predicates Linguistic Migration, in *Proceedings of the EUSFLAT* (2011), pp. 598– 604
14. E. Trillas, A. Sobrino, C. Moraga, On familly resemblances with fuzzy sets, in *Proceeding of the IFSA-EUSFLAT* (2009), pp. 306–311
15. Lord Kelvin (W. Thompson), Lecture on electrical units of measurement, in *Popular Lectures and Addresses*, vol. I (1889), p. 73
16. A.N. Kolmogorov, *Foundations of the Theory of Probability* (Chelsea, New York, 1956)
17. B. De Finetti, Sul significato soggettivo della probabilità. Fundam. Math. **XVII**, 298–329 (1931)

Chapter 3
Antonyms. Negation, and the Fuzzy Case

Can an adult person recognize with the naked eye that John is tall, but not, simultaneously, that Peter is short? It depends, obviously, on the (ostensible) height of both John and Peter, and from what it contextually can mean to be, respectively, a tall and a short person, that is, of perceptively capturing the meanings of these two words when applied to people in some particular context. It also seems, indeed, that people learn the meaning of a word by, simultaneously and by polarity, learning the meaning of one of its opposite words, or antonyms. Antonyms are used in plain language, and usually they are clearly distinguished from negation; it is the case, for instance, of "empty" and "not full", "short" and "not tall", and so on.

The linguistic phenomenon of antonymy is an important one in plain language, and it is (almost) excluded in the artificial ones. It should be pointed out, for instance, that both words "tall" and "short" are in the dictionary, but that their negations, not-tall and not-short, are not in it; hence, if, in this sense, tall and short could be called "linguistic terms", not-tall and not-short are not so, and just mean the exclusion of what can be covered by the initial words tall and short; all that, in the corresponding universe of discourse, is not fully covered by tall, or by short. If negation tries totally to exclude, antonymy only excludes partially. Given a linguistic term P, its antonyms, if existing, are also linguistic terms but its negation is not clearly so. Note that the same word can have different antonyms in different contexts. A first question concerns what the antonyms of P mean, and its negation, once the meaning of P is known.

3.1. Denote by P^a an antonym of P, and by P' its negation not-P; notice that if P can have several antonyms, the negation is unique although it can have different properties in each context. It can be said that antonyms concern plain language, and the negation is more a logical concept expressing, in a given context, what is excluded by P. For their part, antonyms P^a keep some interaction with P. Negation and antonym should not be identified, and for showing how antonym and negation are linked, the following and typical example is illustrative.

E. Trillas, *On the Logos: A Naïve View on Ordinary Reasoning and Fuzzy Logic*,
Studies in Fuzziness and Soft Computing 354, DOI 10.1007/978-3-319-56053-3_3

If P = full, it is P' = not full, and P^a = empty, and, with it, the statement, “If the bottle is empty, then it is not-full”, shows the rule:

$$\text{If } x \text{ is } P^a, \text{ then } x \text{ is } P', \text{ that can be explicated by “If } P^a, \text{ then } P'\text{”},$$

but not reciprocally, because it is not always the case that “If the bottle is not-full, then it is empty”.

Only when P has no antonyms in language is it sometimes taken P' instead of P^a. P is said to be a “regular” word if P^a is not coincidental with P'; otherwise, P is irregular.

3.2. Which are the, in principle different, meanings of P' and P^a, given that of P? Supposing that the quantity $(X, <_P, m_P)$ reflects a meaning of P in a universe X, the problem consists in finding quantities $(X, <_P{}^a, m_P{}^a)$ and $(X, <_{P'}, m_{P'})$ for, respectively, P^a and P' in X.

Because “oppositeness” implies that if “x is less P^a than y”, then “x is more P than y”, and reciprocally, it should be

$$x <_P{}^a y \Leftrightarrow y <_P x \Leftrightarrow x <^{-1}{}_P y,$$

and $<_P{}^a = <^{-1}{}_P$ follows. That is, P and P^a have inverse qualitative meanings in X, and to obtain a measure $m_P{}^a$ from m_P, functions $s: X \to X$ reversing the relation $<_P$ seems to be suitable.

Defining $m_P{}^a = m_P \circ s$, it follows that

(1) $x <_P{}^a y \Leftrightarrow y <_P x \Rightarrow s(x) <_P s(y) \Rightarrow m_P(s(x)) \le m_P(s(y)) \Leftrightarrow m_P{}^a(x) \le m_P{}^a(y)$.
(2) If x is maximal for $<_P{}^a$, it is also maximal for $<^{-1}{}_P$, and minimal for $<_P$. Hence $s(x)$ is maximal for $<_P$. Then $m_P{}^a(x) = m_P(s(x)) = 1$.
(3) x minimal for $<_P{}^a \Rightarrow x$ minimal for $<^{-1}{}_P \Rightarrow x$ maximal for $<_p \Rightarrow s(x)$ minimal for $<_P$. Then $m_P{}^a(x) = 0$.

Thus $m_P{}^a = m_P \circ s$ gives the quantity $(X, <^{-1}{}_P, m_P \circ s)$ reflecting a full meaning of the antonym P^a of P in X. Of course, because each mapping s will give a different measure, it is clear that the antonym depends on the particular form of reversing the graph $(X, <_P)$.

Note that s is actually associated with P; that is, it is more properly denoted by s_P, even if for simplicity it is just written s. For instance, an opposite of P = big in [0, 10] is P^a = small, and, because $<_{big} = \le$, it is $<_{small} = \le^{-1} = \ge$. With the mapping $s(x) = 10 - x$, reversing the relation $<_{big} = \le$, then $m_{small}(x) = m_{big}(10 - x) = 1 - x/10$ would be obtained, provided the linear measure $m_{big}(x) = x/10$ for big were taken. Were it taken $m_{big}(x) = x^2/100$, it would result in $m_{small}(x) = x^2/100 - x/5 + 1$.

It should be pointed out that it is not proven that the only way of obtaining antonyms is through symmetries even if, actually, they seem suitable for reflecting the oppositeness of meaning. Symmetry is generally viewed as a relevant concept in science, and it is always interesting to study if the symmetry shows invariants, analyzing if the symmetry's fixpoints actually correspond with something actually existing in the ground reality. To some extent, symmetries open a door towards a possible geometrical study of antonymy.

A nice application of what has been said lies in the possibility of finding antonyms for words with no antonym in language, through specifying their meaning and naming it. A toy example can be obtained, for instance, with a word P naming the property of being "greater than 6" in [0, 10], as it could be, for instance, the invented word P = "gresix"; it is without an antonym in language. In this case, P is precise and specified by the set (6, 10]; P' is also precise and is specified by [0, 6], with P' naming "between 0 and 6". But, what about P^a? With the same symmetry $s(x) = 10 - x$, it follows that $m_P{}^a(x) = m_P(10 - x) = 1$, if $6 < 10 - x \leq 10$, and zero otherwise. Hence it is $m_P{}^a(x) = 1$ for x in [0, 4), and equal to 0 in [4, 10], that shows a possible solution of P^a by attributing a name to the property of being "strictly smaller than 4".

This example opens a window for extending the idea to words in plain language, passing from this (toy) example in mathematics to true ones in language.

Note that mathematicians don't need to attribute specific words to concepts such as "gresix" is to "greater than 6", inasmuch as these concepts are precise formal ones just expressible by the "primitive" concepts with which they are defined and in this way a saving of words is done; the language of mathematics doesn't always require new words for what appears, because all that is needed comes by if and only if definitions from the axioms. In this way, the language of mathematics economizes unnecessary new words. As Alfred North Whitehead and Bertrand Russell remark in the preface to volume one of their famous book *Principia Mathematica*, if mathematics can be constructed with the help of a few words, plain language needs all of them.

To end with opposites, it is worth remarking that the passing from P to P^a can almost always be seen as an involution; for instance, if small = big^a, it is clear that $(\text{big}^a)^a = (\text{small})^a$ = big. Because it is $<_{(P^a)}{}^a = <^{-1}{}_{P^a} = <_P$, it does not seem bizarre to accept, for opposites, the law of involution, $(P^a)^a = P$. Hence, it should be $m_{(P^a)}{}^a = m_p$, implying $m_P{}^a \ o \ s_P{}^a = (m_P \ o \ s_P) \ o \ s_P{}^a = m_P \ o \ (s_P \ o \ s_P{}^a) = m_P$ thus, and under good conditions, $s_P \ o \ s_P{}^a = \text{id}_X$ can follow; provided these mappings s were bijective, it would imply $s_P{}^a = s_P{}^{-1}$, and if s_P were involutive then it would be but a full symmetry in X, as it is $s(x) = 10 - x$ in [0, 10], verifying $s^2(x) = s(10 - x) = x$, with $s = s^{-1}$, and $s(0) = 10$, $s(10) = 0$.

Thus, opposition can be seen as a symmetry in language, and P is a linguistic fixpoint for such opposition, or a self-opposite word, if its measure satisfies the equation $m_P = m_P{}^a \Leftrightarrow m_P = m_P \ o \ s$, for some symmetry $s \neq \text{id}_X$, and that, provided m_P were invertible, would not exist because, in such case, it would be $s = m_P{}^{-1} \ o \ m_P = \text{id}_X$.

For instance, the former "big" is not self-opposite in [0, 10], because $x/10 = s(x)/10$ implies $s(x) = x$. Anyway, if actually existing, self-opposites are a very rare linguistic phenomenon that only can be thought for words with noninvertible measures, for instance, the word "five" in [0, 10], specified by the singleton $\{5\}$ and for which, with $s_{\text{five}}(x) = 10 - x$, is

$$s_{\text{five}}(5) = 5, \text{ and } m^{a}_{\text{five}}(x) = m_{\text{five}}(10-x) = 1, \text{ if } x = 5; \text{ equal to } 0, \text{ if } x \neq 5;$$

that is, $m_{\text{five}}{}^{a}$ coincides with m_{five}, allowing us to see "five" as an opposite of "five". Anyway, to avoid the character of self-opposite of "five", it suffices to consider a symmetry s such that $s(x) = 5$, for some $x \neq 5$, because it is $m_{\text{five}}{}^{a}(x) = m_{\text{five}}(s(x)) = 1 \Leftrightarrow s(x) = 5$, and then the set $s^{-1}(5)$ would specify, and name, the term five^{a}; if, for instance, it were $s^{-1}(5) = \{7\}$, it would be five^{a} = seven, with "five" losing the former self-opposite's character.

All this shows that if self-opposites are certainly rare, and if even self-opposition can seem a pathology of language, perhaps useful in jokes or poetry, nevertheless, they can exist in, at least, some very particular parts of language; in the end, for instance, the words denoting numbers have a very precise meaning and are often used in language.

3.3. The analysis of the negation P' of P is more complex than that of the opposites P^{a}, because the only thing that can be taken for sure is that $<_P$ is contained in $<_{P'}{}^{-1}$, that, if "x is less P than y", then "y is less P' than x", but without the reciprocal always being sure. That is, in general it can only be presumed

$$<_P \subseteq <_{P'}{}^{-1}, \text{ that is equivalent to } <_{P'} \subseteq <_{P'}{}^{-1}$$

In general, the qualitative meaning of the negation is just a part of the qualitative meaning of the antonyms, even if denying its coincidence cannot be done in all cases. Each antonym P^{a} of P has a larger or equal inverse qualitative meaning than its negation P'; thus, provided P were measurable, it could not be guaranteed that P' is also measurable.

Nevertheless, in the case of coincidence between $<_{P'}$ and $<_P{}^{-1}$ (in which P' is certainly measurable), it is easy to see that the qualitative meaning of the double negation, $(P')' = P''$, coincides with that of P:

$$<_{P''} = <_{P'}{}^{-1} = <_P.$$

Hence a sufficient condition for having P and P'' the same qualitative meaning, is the coincidence of the meaning of P' with the inverse of P. On the contrary, the qualitative meaning of P' is just contained in that of the antonym.

If m_P and $m_{P'}$ are, respectively, measures for the meanings of P and P', from $<_{P'} \subseteq <^{-1}{}_P$ it follows that if $x <_{P'} y$ then $m_{P'}(x) \leq m_{P'}(y)$, and because it is $y <_P x$, then it is also $m_P(y) \leq m_P(x)$; both numerical inequalities coexist when it is $x <_{P'} y$.

Thus, if N: [0, 1] $\rightarrow$ [0, 1] is an order-reversing mapping such that $N(0) = 1$ and $N(1) = 0$, the function $N \, o \, m_P$ verifies:

(1) If $x <_P y$, from $m_P(y) \leq m_P(x)$ follows $N(m_P(x)) \leq N(m_P(y))$.
(2) If x is maximal for $<_{P'}$, it should be minimal for $<_P$, $m_P(x) = 0$, and $N(m_P(x)) = N(0) = 1$.
(3) If x is minimal for $<_{P'}$, it is maximal for $<_P$, $m_P(x) = 1$, and $N(m_P(x)) = N(1) = 0$.

Hence $m_{P'} = N \, o \, m_P$ is a measure for $<_{P'}$, that, if it is $P'' = P$, and taking into account that N actually depends on P, it is better to denote it by N_P because it should be $m_{P''} = m_P$; it is $N_{P'} \, o \, m_{P'} = m_P$, and $N_{P'} \, o \, (N_P \, o \, m_P) = (N_{P'} \, o \, N_P) \, o \, m_P = m_P$, implying, under good conditions, $N_{P'} \, o \, N_P = \mathrm{id}_{[0,\ 1]}$. Then, provided either $N_{P'}$ or N_P were bijective, it would follow that $N_{P'} = {N_P}^{-1}$, and provided it were involutive, it could be seen as a symmetry in [0, 1] whose fixpoints can come from solving $m_{P'} = m_P \Leftrightarrow N_P \, o \, m_P = m_P$ that, newly under good conditions, leads to the nonorder reversing function $N_P = \mathrm{id}_{[0,\ 1]}$, that is not a negation.

The problem of self-negation is similar to that of self-opposition; if it seems that linguistic terms coincidental with their negation cannot exist, nevertheless and because functions N often have fixpoints x different from 0 and 1, $x = N(x)$, a window is open to the existence of self-negating words.

For instance, $N(x) = 1 - x$ has the fixpoint $x = 0.5$, and $N(x) = 1 - x^2$ has the fixpoint coming from the equation $x^2 + x - 1 = 0$, or $x = (\sqrt{5} - 1)/2$, both being order-reversing functions and changing 0 in 1, and 1 in 0. In this way, the words "one half", and the statement "a half of the square root of five minus one", can be seen as self-negating. Notice that both negations are continuous, but the first is involutive, $N(N(x)) = 1 - N(x) = 1 - (1 - x) = x$, and the second is not, $N(N(x)) = 1 - (1 - x^2)^2 = x^2(1 - x^2) \neq x$.

In the case $P = P''$, an involutive function N seems suitable for obtaining a measure for P, and then such N: $[0, 1] \to [0, 1]$, should satisfy the properties:

(1) $x \leq y \Rightarrow N(y) \leq N(y)$
(2) $N(N(x)) = x$ for all x in $[0,\ 1]$
(3) $N(0) = 1$

from which follows $N(1) = N(N(0)) = 0$, $N^{-1} = N$, and also that N is continuous, hence strictly nondecreasing. There are continuous functions N, such as the former $N(x) = 1 - x^2$, verifying 1, 3, and $N(1) = 0$, but not (2); with them, $N \, o \, (N \, o \, m_P)$ cannot coincide with m_P and, hence, are not able to represent those P such that $P'' = P$. For instance, $1 - (1 - (x/10)^2) = x^2/100 \neq x/10$, and under this negation not (not big) cannot coincide with big, but simply not (not big) implies big.

Those negation functions verifying 1, 2, and 3, are called strong negations, and are characterized by the theorem:

- N is a strong negation if and only if there exists an order automorphism Δ of the unit interval, such that $N(x) = \Delta^{-1}(1 - \Delta(x))$, for all x in [0, 1].

Note that just taking $N(x) = 1 - \Delta(x)$, a negation function is also obtained but fails to verify (2) to be involutive. For instance, with $\Delta(x) = x^n$, the family of strong negations $(1 - x^n)^{1/n}$ is obtained, as well as the nonstrong negations $1 - x^n$; with more complex functions Δ, depending on a real parameter $\lambda > -1$, the Sugeno family of strong negations $(1 - x)/(1 + \lambda x)$ is obtained.

That theorem can be easily extended to any closed interval $[a, b]$ of real numbers by just considering that Δ is an order morphism of $[a, b]$; then $s(x) = (a + b) - \Delta(x)$ can serve as a generator of opposites acting in $[a, b]$, and if an involution is preferred, it suffices to take $s(x) = \Delta^{-1}((a + b) - \Delta(x))$. Notice that because it should be $\Delta(a) = a$, and $\Delta(b) = b$, in both cases it should be $s(a) = b$ and $s(b) = a$. A very simple example in the interval [0, 10], with $\Delta = id_{[0, 10]}$, gives the former symmetry $s(x) = 10 - x$.

3.4. How to proceed when P is not directly known by a quantity reflecting its meaning in X, but indirectly through a membership function μ representing the collective/fuzzy set P generates in X?

Suppose that the opposites P^a of P are represented by membership functions μ^a, and the negation P' by μ'. Then, can it be supposed that, for each x in X, there are symmetries $s_x: X \to X$, and negation functions $N_x: [0, 1] \to [0, 1]$, such that $\mu^a(x) = \mu(s_x(x))$, and $\mu'(x) = N_x(\mu(x))$? But, are these formulae actually facilitating an opposite and a negation, respectively? The answer is affirmative for the negation, because

(a) Provided $\mu \leq \sigma$, that is, $\mu(x) \leq \sigma(x)$, for all x in X, it would be $N_x(\sigma(x)) \leq N_x(\mu(x)) \Leftrightarrow \sigma'(x) \leq \mu'(x) \Leftrightarrow \sigma' \leq \mu'$.
(b) $\mu_0'(x) = N_x(\mu_0(x)) = N_x(0) = 1$, for all x in X, or $\mu_0' = \mu_1$, with μ_0 and μ_1 the functions constantly equal to 0 and 1, respectively, that represent the empty subset Ø, and the total subset X.
(c) In the same vein as in (b), it is $\mu_1'(x) = N_x(\mu_1(x)) = N_x(1) = 0$, and thus $\mu_1' = \mu_0$.
(d) Provided all functions N_x were strong negations, then $\mu''(x) = ((\mu')')(x) = N_x(N_x(\mu(x))) = (N_x \, o \, N_x)(\mu(x)) = \mu(x)$, for all x in X, or $\mu'' = \mu$.

Note that this not only holds with a different function N_x for each x in X, but also provided they were the same function for all x in each subset of a partition of X. For instance, a negation of "big" in [0, 10] can be obtained from the partition [0, 10] = [0, 7] U (7, 10], by taking the negations $N_1(x) = 1 - x$ in [0, 7], and $N_2(x) = (1 - x)/(1 + x)$ in (7, 10], and giving:

$$\mu'(x) = 1 - x/10 \text{ if } 0 \leq x \leq 7,$$

and

$$\mu'(x) = (1 - x/10)/(1 + x/10) = 10 - x/10 + x, \text{ if } 7 < x \leq 10.$$

Each N_x is a strong negation, therefore it has a fixpoint $n(x)$, and the family $\{N_x;\ x \text{ in } X\}$ has the curve of fixpoints defined by the function $y = n(x)$; it has a fix curve constituted by the fixpoints of the negations N_x.

For instance, in the last example, because the fixpoint of the first negation function is $n(x) = 0.5$, and that of the second is the positive solution of the equation $1 - x/1 + x = x \Leftrightarrow n(x) = \sqrt{2} - 1$, the corresponding fix curve is

$$n(x) = 0.5, \text{ if } x \text{ is in } [0,\ 7], \text{ and}$$
$$n(x) = \sqrt{2}-1, \text{ if } x \text{ is in } (7,\ 10].$$

Were the negations $N_x(x) = 1 - x/1 + \lambda(x).x$, with the parameter $\lambda(x)$ such that $-1 < \lambda(x)$, the fix curve is given by $1 - x/1 + \lambda(x)x = x$, or $\lambda(x).x^2 + 2\times - 1 = 0$, whose positive solution for $\lambda(x) \neq 0$ (because for $\lambda(x) = 0$ is $x = 0.5$, corresponding to the negation $1 - x$) is

$$n(x) = (1/\lambda(x))\left(\sqrt{(1+\lambda(x)-1)}\right)$$

that, with $\lambda(x) = 1$, for all x, is $n(x) = \sqrt{2} - 1$, as said above. With $\lambda(x) = x$ for all x in [0, 1], the fix curve is

$$n(x) = (1/x)\left(\sqrt{(1+x)} - 1\right), \text{ if } x \text{ in } (0,\ 1], \text{ and } n(0) = 0.5.$$

The fix curve of a family of strong negations is a membership function of a self-negating linguistic label. Looking at the fixpoint $n(x)$ of a single negation as the singleton $\{n(x)\}$, its crisp membership function has a linguistic label whose negation is itself. Self-negation is rare and it can even be seen as a pathological linguistic phenomenon, but it can exist.

A perhaps interesting application of self-negating statements appears in the so-called liar's paradox, in which a (measurable) statement "x is P" is simultaneously viewed as true (T), and false (F); it implies $m_T(\mu_P(x)) = m_F(\mu_P(x)) = m_T(s_T(\mu_P(x)))$. This equality holds if $\mu_P(x) = s_T(\mu_P(x))$, that is, if the degree up to which "x is P" is a fixpoint of the symmetry s_T allowing us to define the measure of false from that of true. In the case of interpreting the paradox as that coming from a statement that is true and not true, the equation is $m_T(\mu_P(x)) = m_{T'}(\mu_P(x)) = N_T(m_T(\mu_P(x)))$, holding if $m_T(\mu_P(x))$ is a fixpoint of the negation N_T.

Note that this is not possible with a precise statement "x is P" whose degree of true is known, because in this case $m_T(\mu_P(x))$ only can be equal to 0 or 1, and these values are not fixpoints of N_T because it is always $N_T(0) = 1$, and $N_T(1) = 0$. From such impossibility comes the paradoxical character of presuming the existence of precise statements simultaneously true and false.

Thus, for actually saying something meaningful, the modalities of the liar's paradox should be stated with an imprecise statement "x is P", or, at least, with one immersed in a context neither allowing us to know its measure of true, nor that of false. It was a statement of this type, found by Bertrand Russell, which collapsed Gottlieb Frege's ideas on the fundamentals of mathematics, and later on led Kurt Gödel and also Alan Turing to their famous theorems that were, and are, among the glorious moments of mathematics in the first third of the twentieth century, and that dismantled David Hilbert's call for a total deductive algorithmization of mathematics. Not even mathematics seems to be possible without guessing.

Concerning the antonyms, given a family of symmetries $\{s_x;$ in $X\}$, the function $\mu^a(x) = \mu(s_x(x))$, for all x in X, can be seen as the membership function of a word with opposite meaning to that labeling μ. As with negation, the mappings s_x can be shared by the x in the several parts of a partition of X, or can be a single one for all x in X.

For instance, if "small" is managed in [0, 10] with the membership function $\mu(x) = 1 - x/10$, and the linear symmetry $s(x) = 10 - x$ is taken, then $\mu^a(x) = \mu(10 - x) = 1 - (10 - x)/10 = x/10$, that is a membership function of "big", and shows a regular antonym of "small", coinciding with its negation if it is given by $N = 1 - \text{Id}$. Anyway, provided it were considered the partition given by the subintervals [0, 6] and (6, 10] and the two symmetries $s(x) = 6 - x$, and $s(x) = 16 - x$ taken at, respectively, each of them, it would be $\mu^a(x) = \mu(6 - x)$ in the first, and $\mu(16 - i)$ in the second; thus, for the antonym "big" it follows that

$$-\mu^a(x) = 1-(6-x)/10 = 0.4 + x/10, \text{ for } x \text{ in } [0, 6], \text{ and}$$
$$-\mu^a(x) = 1-(16-x)/10 = x/10 + 3/5, \text{ for } x \text{ in } (6, 10].$$

Provided the linguistic label P of μ were known, but not any of its antonyms P^a, then a suitable linguistic label should be assigned to μ^a through analyzing, accordingly with P, the characteristics of such membership function in X; for instance, if it resulted in the function μ^a defined by $\mu^a(x) = 0$, for x in [0, 2], and $\mu^a(x) = (x - 1)/8$, for x in (2, 10], some word designing "big after 2" in [0, 10] can be well assigned to P^a.

A way to proceed is not theoretically known for assigning linguistic labels to all the functions $X \to [0, 1]$; there are too many functions. Anyway, in praxis and in each problem it is often contextually possible when several linguistic labels are considered, by comparing the unlabeled membership function with those whose linguistic labels are known. It is the unsolved problem of how to approach linguistically the labeling of a fuzzy set given by a membership function, the problem of "linguistic approximation", in short.

It should be pointed out that the membership functions of negation and antonyms are not independent, but linked by the inequality $\mu^a \le \mu'$, translating that if x is P^a, then x is P', but usually without holding the reverse inequality. Hence, provided the antonym's membership function were constructed thanks to a symmetry s, and that of the negation by a negation function N, it should be checked that $\mu(s(x)) \le N(\mu(x))$ holds for all x in X. This inequality is, in each case, a "condition of coherence" between s and N that, provided it is not verified, would imply that either s or N is not suitable for the problem submitted to a design in fuzzy terms.

For instance, supposing that "big" is managed by its membership function $x/10$ in [0, 10], and that "small" is with $1 - x/10$ (by using the symmetry $s(x) = 10 - x$), to design a membership function of "not big"; those negation functions N only can be used verifying the coherence inequality $1 - x/10 \le N(x/10)$, for all x in [0, 10], that is, those such that $1 - y \le N(y)$, for all y in [0, 1]. Of course, $N(y) = 1 - y$ (with which "small" coincides with "not big"), $N(y) = 1 - y^2$, and $N(x) = (1 - y^2)^{1/2}$ are suitable, but it is not $N(x) = (1 - y)/(1 + y)$. Note that, as is easy to check, the only strong negations in the family $(1 - y)/(1 + \lambda y)$, with $-1 < \lambda$, that are coherent with the symmetry $s(x) = 10 - x$, are those for which it is $-1 < \lambda \leq 0$, as is, for instance, $N(y) = (1 - y)/(1 - 0.4y)$.

Analogously, if what is known is the negation function $N(x) = 1 - x^2$, the symmetry s should verify $s(x)/10 \leq 1 - x^2/100 \Leftrightarrow s(x) \leq 10 - x^2/10$, with which it is clear that the symmetry $10 - x$ is coherent with the (not strong) negation $1 - x^2$, and it is easy to prove that it is also consistent with all the strong negations $(10 - x)/(10 + \lambda x)$, $-1 < \lambda$.

3.5. All that has been shown in this chapter illustrates why the design of fuzzy systems should be carefully done according to the theoretic mathematical armamentarium of fuzzy sets. For instance, if both the negation and an antonym of a linguistic label P intervene in the problem, the designer should be very careful, first in designing the membership function μ_P according to as much information as possible that is available to him on the qualitative meaning $(X, <_P)$, with plausible hypotheses on the shape of μ_P for the problem in course, and, after designing μ_P, deciding which coherent symmetry and negation function will be taken for building up the membership functions $\mu_{P'}$ and $\mu_P{}^a$. In addition, and concerning the negation function, the designer should previously capture whether it allows identifying P'' and P, and, on the other hand if P follows from P'', or P'' follows from P, or P and P'' were not comparable; provided it were not $P'' = P$, the designer cannot chose a strong negation, but one respecting the existing relation between P and P". For instance, because with the negation $N(x) = 1 - x^2$ is $N(N(x)) = N(1 - x^2) = 1 - (1 - x^2)^2 \leq x$, it can only be used in those cases in which it can be presumed that $\mu_{P''} \leq \mu_P$, that P follows from P''.

All that implies that the designer of fuzzy systems, as well as the practitioner of fuzzy logic, neither designs a system in a blind form, nor tries to do it without some prior and theoretical knowledge of the theory of fuzzy sets; usually, confidence in the presumed guide done by a practical manual can mean both blindness for a correct design, and ignorance of a very elemental knowledge of what fuzzy sets and their algebra are. Let's repeat that because of its linking with plain language, everything in fuzzy logic is not only a matter of degree but, because everything in it is context-dependent and purpose-driven, is also a matter of design. As said before, a fuzzy set is a kind of cloudy entity that only can be well considered through its contextual crisp states, the membership functions.

3.6. A comment, based on some examples, on the negation of an antonym and the antonym of a negation is still in order. If P = tall, and P^a = short, we have P' = not tall; hence, $(P^a)'$ = not short, but $(P')^a$ = (not tall)^a needs to be defined because "not tall" is not a linguistic term; it can be easily accepted, in this case, that it is (not tall)^a = not short, that is, accepting the rule/definition $(P')^a = (P^a)'$, showing a commuting between antonym and negation. Notwithstanding, with P = employed, we have P' = not employed, and P^a = unemployed, and then (not employed)^a does not seem to be clearly identifiable with not unemployed. Hence the commuting between "not" and "antonym" does not seem to be always clear enough in language; it seems to be not formally decidable as a general rule. Let's see what can show the corresponding membership functions in such a respect.

The membership function of $(P^{a})'$ is $\mu_{(P^{a})}{}'(x) = N_{P^{a}}(\mu_P(S_P(x)))$ and that of $(P')^{a}$ is $\mu_{(P')}{}^{a}(x) = \mu_{P'}(S_{P'}(x)) = N_P(\mu(S_{P'}(x)))$; hence, without the equalities $N_{P^{a}}$ and $s_P = s_{P'}$ the coincidence between both membership functions is not possible in general. It depends on the particular forms in which both the negation and the antonym should be contextually understood. The commutation between negation and antonym depends on which functions s and N is the available contextual information on P' and P^{a} with which the designer is allowed to count.

References

1. E. Trillas, A. de Soto, On antonym and negate in fuzzy logic. Int. J. Intell. Syst. **14**(3), 295–303 (1999)
2. E. Trillas, C. Moraga, S. Guadarrama, S. Cubillo, E. Castiñeira, Computing with antonyms, in *Forging New Frontiers*, vol. I, ed. by M. Nikravesh et al. (Springer, New York, 2007), pp. 133–153
3. E. Trillas, Sobre funciones de negación en la teoría de conjuntos difusos. Stochastica **III**(1), 47–60 (1979)
4. E. Trillas, A. Pradera,, On non-functional fuzzy connectives. The case of negations, in *Proceedings of the ESTYLF* (2002) pp. 527–532
5. S.V. Ovchinnikov, General negations in fuzzy set theory. J. Math. Anal. Appl. **92**(1), 234–239 (1983)
6. R. Lowen, On fuzzy complements. Inf. Sci. **14**, 107–113 (1978)
7. L.R. Horn, *A Natural History of Negation* (CSLI Publications, Stanford, 2001)
8. A. Lehrer, K. Lehrer, Antonymy. Linguis. Philos. **5**, 483–490 (1982)
9. B. Callejas Bedregal, On fuzzy negation and automorphisms. Anais do CNMAC **21,** 1125–1130 (2009)
10. F. Esteva, E. Trillas, X. Domingo, Weak and strong negations for fuzzy set theory, in *Proceedings of the Symposium on Multiple-Valued Logic* (1981), pp. 23–27
11. E. Trillas, C. Alsina, J. Jacas, On contradiction in fuzzy logic. Soft. Comput. **3**, 197–199 (1999)

Chapter 4
"And" and "Or" in Language: The Case with Fuzzy Sets

Perhaps the simplest form of constructing a statement with two constituent parts is joining them by using the linguistic connectives *and* and *or*, as done, for instance, with the elemental parts "x is P", and "y is Q", for obtaining either "x is P *and* y is Q", and "x is P *or* y is Q", with P and Q acting, respectively, in universes X and Y. It shows that, initially, the particle-words *and,* and *or,* operate in the universe $X\ [P] \times Y\ [Q]$, and, hence, their behavior/action should be searched for in it.

In plain language, and contrary to the artificial ones, these particles are not "logical constants", but are endowed with a meaning that should be specified in each context; the meanings of "P *and* Q" and "P *or* Q", not only depend on the meanings of P and Q, but also on those of *and* and *or*. A first question is, thus, how to capture the meanings of "P *and* Q" and "P *or* Q".

4.1. How can the action of *and* be captured? In the classical setting, in which the parts joined by *and* are precise and hence specified by subsets **P** and **Q** in the respective universes of discourse X and Y, the "new" predicative word "P and Q": $= P \,\&\, Q = P \cdot Q$, in $X \times Y$, is understood as "x is P and y is Q" $\Leftrightarrow$ (x, y) is P & Q, and is specified in X U Y by the intersection of its subsets **P** and **Q**.

In general, once the qualitative meanings of P in X and Q in Y are known, it should be If x_1 is less P than x_2 and y_1 is less Q than y_2, then (x_1, x_2) is less P & Q than (y_1, y_2), shortened by

$$<_P \times <_Q \subseteq <_{P\&Q},$$

And showing that $<_{P\&Q}$ is not empty provided the Cartesian product $<_P \times <_Q$, were not so, that P & Q would be measurable. If it can be accepted that "(x, y) is P & Q" implies "x is P" and also "y is Q", then both would be equivalent, that is, $<_{P\&Q} = <_P \times <_Q$, as happens in the precise case.

E. Trillas, *On the Logos: A Naïve View on Ordinary Reasoning and Fuzzy Logic*,
Studies in Fuzziness and Soft Computing 354, DOI 10.1007/978-3-319-56053-3_4

Concerning the specification of measures $m_{P\&Q}$ obtained once measures m_P and m_Q are specified, it should be noticed that a priori nothing guarantees the existence of a two-variable function A: $[0, 1] \times [0, 1] \to [0, 1]$ (and-function), such that $m_{P\&Q} = A \text{ o } (m_P \times m_Q)$, that is, that the measure of P & Q should be decomposable, or functionally expressible, by means of the measures of P and Q. Nevertheless, in the case of nondecomposability, the only thing that can be done is directly checking that $m_{P\&Q}$ verifies, in the graph $(X \times Y, <_{P\&Q})$, the three axioms for a measure, something not often easy to do. In addition, proving that a given measure $m_{P\&Q}$ is nondecomposable only can be done directly by finding, at least, one pair (x, y) with two different values under it. It is for reasons of this kind that decomposable measures are usually preferred in praxis; they save the designer from such checking, even if she or he cannot forget to consciously assume that decomposability is but a hypothesis.

Hence finding those functions A is relevant for the praxis and, in principle, provided it were $<_P x <_Q = <_{P\&Q}$, the designer should necessarily verify the following properties.

(1) Because $(x_1, x_2) <_{P\&Q} (y_1, y_2)$, is equivalent to $x_1 <_P x_2$, and $y_1 <_Q y_2$, then it is $m_P(x_1) \leq m_P(x_2)$, and $m_Q(y_1) \leq m_Q(y_2)$. Hence it suffices a nondecreasing function A in its two variables to have

$$A(m_P(x_1),\ m_Q(y_1)) \leq A(m_P(x_2), m_Q(y_2)) \Leftrightarrow m_{P\&Q}(x_1,\ x_2) \leq m_{P\&Q}(y_1,\ y_2);$$

that is, under this presumption, $m_{P\&Q}$ verifies the first axiom of a measure.

(2) Provided (x, y) is maximal for $<_{P\&Q}$ implies that x is maximal for $<_P$, or y for $<_Q$, and that $A(1, x) = A(x, 1) = 1$, then $m_{P\&Q}(x, y) = A(1, x)$ (or $A(x, 1)) = 1$. Notice that if there is coincidence between both maximal characters, it suffices the property $A(1, 1) = 1$.

(3) Provided (x, y) is minimal for $<_{P\&Q}$ implies x is minimal for $<_P$, or y for $<_Q$, and that $A(0, x) = A(x, 0) = 0$, then $m_{P\&Q}(x, y) = 0$. With coincidence between the minimal characters, it suffices the property $A(0, 0) = 0$.

Consequently, under good enough conditions, a function A suffices, nondecreasing in each variable, taking the value 1 at (1, 1), and 0 at (0, 0), to count with the measure $m_{P\&Q} = A \text{ o } (m_P \times m_Q)$ for the meaning of P & Q.

An analogous reasoning can be made for representing a more complex statement such as, '(x is P *and* y is Q) *and* z is R), with $m_{(P\&Q)\&R} = A_2(A_1(m_P \times m_Q), m_R)$, for A_1 corresponding to P & Q, and A_2 to R & (P & Q) that, in praxis, are always considered to be coincidental. This cannot always be presumed; for instance, in a large statement with several appearances of *and*, it should be previously checked if all of them can be represented, or not, by the same function A.

Note that one of such functions is $A(x, y) = \min(x, y)$, a commutative and associative operation easily translated into n-variables one by one, for instance, $A(x, y, z) = \min(x, y, z)$, without taking care of the way in which x, y, z, are ordered. It is not always possible with a nonassociative two-variable function A, for instance, with the arithmetical mean $(x + y)/2$ that, nevertheless, is commutative, but is not the pondered mean $(3x + 2y)/5$.

A remark is still in order for the case $P = Q$, (x, y) is P & P, in which, and in principle, if "x is P *and* y is P", then "x is P", and also "y is P", but without knowing if P & P coincides with P or not; in the first case, function A should verify $A(x, x) = x$, for all x in [0, 1] and not only for $x = 0$ and $x = 1$, but in the second functions A such that, generally, $A(x, x) \neq x$ is required. Notice that presuming $A(x, x) = x$, implies to presume the meaning of P & P coincides with that of P, something that, clear in the precise meanings of mathematical terms, is not so clear with the imprecise terms of plain language. For instance, if we can accept that "7 is prime *and* 7 is prime", shortened as "7 is prime *and* prime", means nothing else than "7 is prime", it is not so with "Richard is crazy *and* crazy", because in this kind of plain language's statements the *and* seems to be used for reinforcing the adjective to, perhaps, the meaning of "very crazy".

Thus, functions A should be selected according to the properties that can be assumed for the actually used linguistic *and*. For instance, selecting a commutative A means to accept that P & Q = Q & P, and accepting an associative A corresponds to accepting (P & Q) & R = P & (Q & R), like accepting an involutive one is to accept P & P = P; that is, the selection of A cannot be blindly done without checking, in each case, if its algebraic properties are effectively verified in what is both linguistically expressed and tried to be represented by A. On the contrary, such properties will be imposed on language, and it can be an artificial supposition affecting the solution of the problem under consideration. Selecting A is a matter of design needing to take into account the language's reality in the considered piece of it. The same properties cannot be presumed and assigned to the language of mathematics as to the language describing something in the world. For instance, if what describes the statement, "7 is prime and less than 11" clearly leads it to commute, what the statement, "He started crying, and entered the room" describes, does not lead immediately to application of the commutative law to its possible symbolic representations; doing it implies a risk the designer should be, at least, able to evaluate. For instance, in the statements, "He is Italian and tall, and he started crying and entered the room", and "He is Italian and tall, and he entered the room and started crying", in principle the several "and" particles cannot be supposed to keep the same properties; different and-functions should be selected for representing them.

4.2. What about the linguistic *or*? That is, how can the action of "P or Q" (in symbols, $P + Q$) through the elemental statements "x is P *or* y is Q", shortened by "(x, y) is $(P + Q)$", the meaning of $P + Q$, be captured and symbolically represented?

In this case what can be presumed is just $(<_P \cup <_Q) \subseteq <_{P+Q}$, which can be viewed as coming from "if x is P, then x is P or Q" whatever it can be Q, and allows us to state that $P + Q$ has a qualitative meaning provided at least P or Q were to have a qualitative meaning. That is, one of $<_P$, or $<_Q$, is not empty; $P + Q$ is measurable provided P or Q were to be measurable.

It cannot always be presumed that $P + P$ coincides with P, as in the classical case with precise words represented by sets, and because it is $\mathbf{P} \cup \mathbf{P} = \mathbf{P}$. For instance, if the "or" is exclusive, that is, if by $P + Q$ it is understood as "either P, or Q", that can be interpreted by "P or Q, but not both", then it can mean nothing. It happens, for instance, in the classical crisp case, where $(\mathbf{P} \cup \mathbf{P}) \cap (\mathbf{P} \cap \mathbf{P})^c = \mathbf{P} \cap \emptyset = \emptyset$. Of course, $(P + P) \cdot (P \cdot P)'$, should be interpreted in each case, and it can be measurable or meaningless because, even provided P were measurable and $P + P$ is not meaningless, $(P \cdot P)'$ could be so. As has been said, the designer of a representation of $P + Q$ or $P \cdot Q$ should be able to check how, in the used piece of language and for the current problem, *and* and *or* are used.

As in the case of $P \cdot Q$, to find a measure m_{P+Q}, decomposable with respect to m_P and m_Q, and once supposed that $(<_P \cup <_Q)$ coincides with $<_{P+Q}$, an or-function, O: $[0, 1] \times [0, 1] \to [0, 1]$, such that $m_{P+Q} = O \circ (m_P \times m_Q)$, should be found. For it,

(1) Because "x_1 is less P than x_2" or "y_1 is less Q than y_2" is equivalent to "(x_1, x_2) is less $P + Q$" than (y_1, y_2)', then

$$m_P(x_1) \leq m_P(x_2) \text{ or } m_Q(y_1) \leq m_Q(y_2) \text{ implies } m_{P+Q}(x_1, x_2) \leq m_{P+Q}(y_1, y_2).$$

Provided O is nondecreasing in both variables, it would follow that $m_{P+Q}(x, y) = O(m_P(x), m_Q(y))$ verifies the first axiom of a measure for the meaning of $P + Q$.

(2) Provided the maximal character of (x, y) for $<_{P+Q}$ were to coincide with x is maximal for $<_P$, and y maximal for $<_Q$, then $m_{P+Q}(x, y) = O(1, 1)$, and, hence, provided the function O verified $O(1, 1) = 1$, it would follow that $m_{P+Q}(x, y) = 1$. Provided only one of x or y were maximal for its respective qualitative meaning, then it should be $O(1, y) = 1$, or $O(x, 1) = 1$, that is, 1 absorbent for function O.

(3) Analogously, with the preserving character of minimality for (x, y), it follows that $m_{P+Q}(x, y) = O(0, 0)$, and, hence, provided $O(0, 0) = 0$, it would follow that $m_{P+Q}(x, y) = 0$. Provided only one of x or y were minimal for its respective qualitative meaning, then it should be $O(0, y) = 0$, or $O(x, 0) = 0$.

Consequently, supposing that O is nondecreasing in both variables verifies $O(0, 0) = 0$, $O(1, 1) = 1$, and under good conditions for maximals and minimals, $m_{P+Q} = O \circ (m_P \times m_Q)$ is a measure for $P + Q$.

Note that or-functions O show, under the above conditions, the same axioms as and-functions A, and hence in praxis they should be distinguished by some additional property inasmuch as they could instead imply the identity $P + Q = P \cdot Q$ that, at least if $P \neq Q$, sounds very rare. In the classical crisp case, it corresponds with the identity $\mathbf{P} \cup \mathbf{Q} = \mathbf{P} \cap \mathbf{Q}$ equivalent to $\mathbf{P} = \mathbf{Q}$, because it is $\mathbf{P} \subseteq \mathbf{P} \cup \mathbf{Q} = \mathbf{P} \cap \mathbf{Q} \subseteq \mathbf{P}$, and $\mathbf{Q} \subseteq \mathbf{P} \cup \mathbf{Q} = \mathbf{P} \cap \mathbf{Q} \subseteq \mathbf{Q}$, an equivalence that, nevertheless and when P or Q are not precise, is not so clear in general.

To obtain such a distinction between A and O, let's assume that the words P and Q verify

$$P \cdot Q \text{ implies } P + Q,$$

leading to $m_{P \cdot Q} \leq m_{P + Q}$, and hence to the coherence's condition:

$$A \leq O, \text{ that is } A(x, y) \leq O(x, y), \text{ for all } x, y \text{ in } [0, 1],$$

with the equality only acceptable in a pathological case. This shows that if $A = \min$, it should be $\min < O$, for instance, $O = \max$; and that if $O = \max$, then it is $A < \max$, such as $A =$ product. In principle, there is an enormous amount of and-functions and or-functions able, respectively, to represent the measures of the used linguistic conjunction and disjunction, that is, they can show very different full meanings.

4.3. Let's now consider the case in which P and Q are known by means of their corresponding designed membership functions μ_P and μ_Q, and what is tried to be obtained directly are the membership functions $\mu_{P \cdot Q}$ and $\mu_{P + Q}$ for, respectively, the linguistic labels "P and Q", and "P or Q". Suppose an and-function A exists, and an or-function O ($A < O$), with which it is $\mu_{P \cdot Q}(x, y) = A(\mu_P(x), \mu_Q(y))$ and $\mu_{P + Q}(x, y) = O(\mu_P(x), \mu_Q(y))$, for all x, y in X. Then,

- The commutative character of the conjunction and the disjunction, $P \cdot Q = Q \cdot P$, $P + Q = Q + P$, requires $A(x, y) = A(y, x)$, and $O(x, y) = O(y, x)$, for all x, y in $[0, 1]$; that is, A and O should be commutative operations.
- The involutive characters, $P \cdot P = P$, $P + P = P$, require $A(x, x) = x$, and $O(x, x) = x$, for all x in X; that is, A and O should be involutive operations.
- The associative characters, $P \cdot (Q \cdot R) = (P \cdot Q) \cdot R$, $P + (Q + R) = (P + Q) + R$, require that A and O are associative operations, that is, verify $A(x, A(y, z)) = A(A(x, y), z)$, $O(x, O(y, z)) = O(O(x, y), z)$, for all x, y, z in $[0, 1]$.
- The distributive characters, $P \cdot (Q + R) = (P \cdot Q) + (P \cdot R)$, $P + (Q \cdot R) = (P + Q) \cdot (P + R)$, require, respectively, the properties $A(x, O(y, z)) = O(A(x, y), A(x, z))$, and $O(x, A(y, z)) = A(O(x, y), O(x, z))$, for all x, y, z in $[0, 1]$; that is, A is a distributive operation for O, and O is such for A. It is the case, for instance, if it is $A = \min$, and $O = \max$, but not if it is $A =$ prod and $O = \max$.

And so on; for instance, if it is $P + P'$ such that its measure is one, $m_{P + P'}(x, y) = O(m_P(x), m_{P'}(y)) = 1$, provided the measure of P' were expressible by a negation function N, it would be $O(m_P(x), N(m_P(x))) = 1$, giving the possibility of finding O and N by solving the numerical functional equation

$$O(a, N(a)) = 1, \text{ for all } a \text{ in } [0, 1],$$

and provided it were possible by benefiting from the properties O and N can have. Analogously, if $P \cdot P'$ has measure zero, $m_{P \cdot P'}(x, y) = A(m_P(x), N(m_P(x))) = 0$ [*], then A and N can be eventually found through solving the numerical functional equation

$$A(a, N(a)) = 0 \text{ for all } a \text{ in } [0, 1].$$

That is, in the praxis, and-functions, or-functions, and negation functions, they can be eventually found through solving some functional equations; for such a goal, the continuity of such functions is a good help as illustrated further on, and apart from the usual necessity that the resulting functions should be continuous as a counterpart for the predicates’ flexibility. In addition and for instance, from solving the last equation solutions also follow for the case in which $P \cdot P^a$ has measure zero, inasmuch as $m_P{}^a(x) \le m_{P'}(x)$, from [*] it follows that $A(m_P(x), m_P(s(x)) = 0$.

All this shows, in a new light, the relevance of carefully designing, in accordance with how things actually are, the measures intervening in linguistic descriptions; in the end, measures are but membership functions even if, as said before, such functions are not always measures, but just approximations to them, whose continuity is important for solving the corresponding functional equations, and can come from the flexibility of the considered imprecise words.

4.4. What has been presented in both Chap. 3, and the last Sects. 4.1–4.3, opens a window to consider what can be understood by a (primitive) algebra of fuzzy sets, namely to establish a general enough algebra of fuzzy sets allowing its particularization in each concrete problem. That is, defining among fuzzy sets an algebraic structure just endowed with a minimal number of axioms, not only necessary for counting with an operative general view of fuzzy sets in relation to its linguistic labels, in short with language, but for having the possibility of designing and effectively computing with fuzzy sets. In fact, for constructing a theory of “computing with words”, a calculus is needed in each context. For such a goal it is required not to consider the linguistic collectives/fuzzy sets in themselves, but all their potential states/membership functions, that is, the functions in the set $[0, 1]^X$ of all functions $X \to [0, 1]$, by counting with the important difference that each bivalued function $f: X \to \{0, 1\}$ actually and univocally represents the single crisp subset $f^{-1}(1)$ of X, something without a parallel with fuzzy sets and the functions in $[0, 1]^X$ (except for those in its subset $\{0, 1\}^X$).

Such an algebraic structure should be established, after defining a suitable ordering in $[0, 1]^X$, with two binary operations, $(\cdot)$ and $(+)$ and a unary one $(')$ representing, respectively, the “and”, “or”, and “not” of language where, as said before, the validity of the laws imposed to such operations should be checked in each case. For this reason a structure with very few laws is important; without some laws there is no mathematical structure able to allow a calculus; they are necessary for developing the consequences from its acceptance, and to count with a solid base for computation, reducible to the classical case when the represented words are precise. What has been said offers a way to establish such a general structure.

Hence, for instance, the ordering defined among the functions in $[0, 1]^X$ should reduce to the ordering among subsets or, equivalently, to the ordering among its characteristic functions, namely $\mathbf{P} \subseteq \mathbf{Q} \Leftrightarrow \mu_P(x) \leq \mu_Q(x)$, for all x in X. Thus, the order chosen for membership functions in $[0, 1]^X$ can be the pointwise one:

$$\mu \leq \sigma \Leftrightarrow \mu(x) \leq \sigma(x), \text{ for all } x \text{ in } X,$$

under which the minimum is the function $\mu_0(x) = 0$ for all x in X, the maximum is $\mu_1(x) = 1$ for all x in X, and it is $\mu_0 \leq \sigma \leq \mu_1$ for all σ in $[0, 1]^X$; functions σ appear under such ordering in the interval of functions $[\mu_0, \mu_1]$ between, respectively, the empty Ø and the total set X. The functions $\mu_r(x) = r$, r in $[0, 1]$, for all x in X, are the constant functions reduced in the classical case to only those with $r = 0$ and $r = 1$. Additionally, it is $\mu = \sigma \Leftrightarrow \mu \leq \sigma$ and $\sigma \leq \mu$. With this:

- A basic algebra of fuzzy sets (BAF) is a quintet $([0, 1]^X, \leq; \cdot, +; ')$ satisfying the laws, or axioms:

  (1) For negation: $\mu \leq \sigma \Rightarrow \sigma' \leq \mu'$; $(\mu_1)' = \mu_0$; $(\mu_0)' = \mu_1$.
  (2) For conjunction: $\mu \leq \sigma \Rightarrow \mu \cdot \alpha \leq \sigma \cdot \alpha$, and $\alpha \cdot \mu \leq \alpha \cdot \sigma$, for all α in $[0, 1]^X$; $\mu \cdot \mu_1 = \mu_1 \cdot \mu = \mu$, for all μ in $[0, 1]^X$.
  (3) For disjunction: $\mu \leq \sigma \Rightarrow \mu + \alpha \leq \sigma + \alpha$, and $\alpha + \mu \leq \alpha + \sigma$, for all α in $[0, 1]^X$; $\mu + \mu_0 = \mu_0 + \mu = \mu$, for all μ in $[0, 1]^X$.
  (4) Of coherence: Provided μ and σ belong to $\{0, 1\}^X$, then it would be $\mu \cdot \sigma = \min(\mu, \sigma)$, $\mu + \sigma = \max(\mu, \sigma)$, $\mu' = 1 - \mu$ $(\sigma' = 1 - \sigma)$, all of them belonging to $\{0, 1\}^X$.

Note that such algebras are neither presumed to verify all the laws classically supposed between sets, nor even those that are usually supposed between fuzzy sets. A BAF is but a "formal skeleton" to which other laws could be added when suitable, such as when searching for operations allowing the satisfaction of Aristotle's principles of noncontradiction, $\mu \cdot \mu' = \mu_0$, of excluded-middle, $\mu + \mu' = \mu_1$, or the commutative law $\mu \cdot \sigma = \sigma \cdot \mu$, among others.

Anyway, a few consequences follow from the small number of laws a BAF verifies.

(a) Axiom 4 is independent of axioms 1, 2, and 3.
In fact, defining $\mu^*(x) = 1 - \mu(1 - x)$, for μ in $[0, 1]^{[0, 1]}$, and all x in $[0, 1]$, it is easy to check that μ^* verifies 1, but if μ is the membership function of $[0, 0.4]$, then μ^* is the membership function of $[0.6, 1]$, not coincidental with the complement $(0.4, 1]$ of $[0, 0.4]$. Hence μ^* cannot be taken as a (coherent) negation of μ.
(b) $\mu \cdot \sigma \leq \min(\mu, \sigma) \leq \max(\mu, \sigma) \leq \mu + \sigma$.
In fact, from $\mu \leq \mu_1$, it follows that $\mu \cdot \sigma \leq \mu_1 \cdot \sigma = \sigma$, and from $\sigma \leq \mu_1$, it analogously follows that $\mu \cdot \sigma \leq \mu$; hence, $\mu \cdot \sigma \leq \min(\mu, \sigma) \leq \max(\mu, \sigma)$.

From $\mu_0 \le \mu$ follows that $\sigma = \mu_0 + \sigma \le \mu + \sigma$, and from $\mu_0 \le \sigma$ it follows that $\mu \le \mu + \sigma$; that is, max $(\mu, \sigma) \le \mu + \sigma$.

(c) $\mu \cdot \mu_0 = \mu_0 \cdot \mu = \mu_0$, and $\mu + \mu_1 = \mu_1 + \mu = \mu_1$.
In fact, the first follows from $\mu_0 \le \mu \cdot \mu_0 \le \mu_0$, and the second from $\mu_1 \le \mu + \mu_1 \le \mu_1$.

(d) It is obvious that the only case in which $([0, 1]^X, \le; \cdot, +; ')$ can be a lattice with negation, is with $\cdot$ = min, and + = max.
In this case, if (') is a strong negation, that is, verifies $(\mu')' = \mu$, for all functions μ, it is easy to prove that $(\mu + \sigma)' = \mu' \cdot \sigma'$, and $(\mu \cdot \sigma)' = \mu' + \sigma'$ hold, the laws of duality. Hence, in this case, disjunction and conjunction are not independent, but (through the negation) one depends on the other, and the BFA is a De Morgan algebra.

(e) With an "abstract" partially ordered set Ω with a maximum 1 and a minimum 0, instead of $[0, 1]^X$, and keeping within its elements the former axioms 1, 2, and 3 of a BAF, both ortholattices and De Morgan algebras are instances of such new and abstract algebraic structures, and Boolean algebras in particular; in them, the role of $\{0, 1\}^X$ in Axiom (4) is played by the subset $\{0, 1\}$. Nevertheless, no BAF on $[0, 1]^X$ can have all the properties of a Boolean algebra; indeed, because such a case is that of a lattice, it should be $\cdot$ = min, + = max, and for no negation (') it is possible to have $\mu_r \cdot \mu_r' = \mu_0$, for all r in [0, 1], because min $(\mu_r(x), \mu_r'(x)) = 0$ for all x in X implies $r = 0$. There is no way, under the BAF's axioms, of endowing all the membership functions of fuzzy sets with either the Boolean structure of the characteristic functions of sets, nor with the weaker of an ortholattice. On the contrary, provided $[0, 1]^X$ were endowed with a Boolean structure, it would follow, through Stone's characterization theorem of Boolean algebras as algebras of sets, the very surprising fact of an isomorphism existing between membership functions and characteristic functions, with which fuzzy sets were not, actually, substantially different from sets.
In sum, the BAF's structure seems to be general enough for representing the states of fuzzy sets and another, eventually different, primitive axiomatic for $[0, 1]^X$, should be able to deduce such a result.

(f) Concerning the validity of the laws of duality, holding, as said, within the lattice structure, they don't hold in all BAF, even if they can hold in some particular cases.
For instance, were the connectives decomposable by a triplet (T, S, N), where S is the N-dual of T, and N is strong, that is, $S(x, y) = N(T(N(x), N(y))$, and $N^2 = N$, then it would obviously hold that $(\mu + \sigma)' = \mu' \cdot \sigma'$, and because it is also $S(N(x), N(y)) = N(T(x, y))$, it would also hold that $(\mu \cdot \sigma)' = \mu' + \sigma'$. Notwithstanding, with T = product, S = max, and $N = 1 -$ id, it is $1 - x \cdot y \ne \max(1 - x, 1 - y) = 1 - \min(x, y) \Leftrightarrow x \cdot y \ne \min(x, y)$, and hence the laws of duality are not valid. The laws of duality do not generally hold in a BAF, and their validity depends on the particular expression of the connectives.

Anyway, in a BAF with $+ = \max$, and regardless of the conjunction, the law of semi-duality $\mu' + \sigma' \leq (\mu \cdot \sigma)'$ holds, and with $\cdot = \min$ and regardless of the disjunction, the other semi-duality law $(\mu + \sigma)' \leq \mu' \cdot \sigma'$ holds. For instance, if $+ = \max$, and because it is $\mu \leq \mu + \sigma$, and $\sigma \leq \mu + \sigma$, it follows that $(\mu + \sigma)' \leq \mu'$ and $(\mu + \sigma)' \leq \sigma'$, from which it follows that $(\mu + \sigma)' = (\mu + \sigma)' + (\mu + \sigma)' \leq \mu' \cdot \sigma'$. With $\cdot = \min$, it can proceed analogously.
Of course, if $+ = \max$, and $\cdot = \min$, both inequalities jointly hold but are reduced to equalities. In general, conjunction and disjunction are, nevertheless, independent operations.

(g) The conjunction $(\cdot)$ is involutive, $\mu \cdot \mu = \mu$, if and only if $\cdot = \min$, and the disjunction $(+)$ is involutive, $\mu + \mu = \mu$, if and only if $+ = \max$.
That these two operations are involutive is well known, but the reciprocal also holds. Let's suppose that $(\cdot)$ is involutive; that is, $\mu \cdot \mu = \mu$ holds, for all functions μ. Because it is $\min(\mu, \sigma) \cdot \min(\mu, \sigma) = \min(\mu, \sigma)$, and $\min(\mu, \sigma) \leq \mu$, $\min(\mu, \sigma) \leq \sigma$, it follows that $\min(\mu, \sigma) \cdot \min(\mu, \sigma) \leq \mu \cdot \sigma$, and $\min(\mu, \sigma) \leq \mu \cdot \sigma$; that is, $\min(\mu, \sigma) = \mu \cdot \sigma$. A similar proof applied to max and + shows $\max = +$.

(h) All BFA verify Kleene's law $\mu \cdot \mu' \leq \sigma + \sigma'$, for all μ and σ in $[0, 1]^X$, that reduces to $\mu_0 \leq \mu_1$, if μ and σ are in $\{0, 1\}^X$, or the BAF verifies the Aristotelian principles $\mu \cdot \mu' = \mu_0$, or $\mu + \mu' = \mu_1$.
Because at each x it is either $\mu(x) \leq \sigma(x)$, or $\sigma(x) \leq \mu(x)$, in the first case, it is $(\mu \cdot \mu')(x) \leq \min(\mu(x), \mu'(x)) \leq \mu(x) \leq \sigma(x) \leq \max(\sigma(x), \sigma'(x)) \leq (\sigma + \sigma')(x)$. In the second it analogously follows the same result, and hence Kleene's law is proven for all BAF.

(i) Concerning the absorption laws, $\mu \cdot (\mu + \sigma) = \mu$, and $\mu + (\mu \cdot \sigma) = \mu$, for all μ and σ in $[0, 1]^X$, the first holds if and only if $\cdot = \min$, and the second if and only if $+ = \max$.
That these two formulae hold with, respectively, min and max, is evident. To prove the reciprocal, by supposing that the first holds, just taking $\sigma = \mu_0$, it follows that $\mu \cdot \mu = \mu$, for all μ, implies $\cdot = \min$. Analogously, and by taking in the second $\sigma = \mu_1$, it follows that $\mu + \mu = \mu$, for all μ, which leads to $+ = \max$.

(j) Provided the operations of a BAF were decomposable by numerical functions $F, G: [0, 1] \times [0, 1] \to [0, 1]$, for conjunction and disjunction, respectively, and $N: [0, 1] \to [0, 1]$, for negation, which properties enjoy these three functions? Obviously, such properties are the following.

    (a) Properties of N: $x \leq y \Rightarrow N(y) \leq N(x)$; $N(0) = 1$; $N(1) = 0$.
    (b) Properties of F: $x \leq y \Rightarrow F(x, z) \leq F(y, z)$, $F(z, x) \leq F(z, y)$, for all z in $[0, 1]$; and $F(1, x) = F(x, 1) = x$.
    (c) Properties of G: $x \leq y \Rightarrow G(x, z) \leq G(y, z)$, $G(z, x) \leq G(z, y)$, for all z in $[0, 1]$; and $G(0, x) = G(x, 0) = x$.
    (d) Coherence: If x, y belong to $\{0, 1\}$, then $F(x, y) = \min(x, y)$, $G(x, y) = \max(x, y)$, and $N(x) = 1 - x$.

Hence ([0, 1], $\leq$; F, G; N) inherits the corresponding BAF structure with the set {0, 1} playing the role of $\{0, 1\}^X$; for instance, functions F and G verify $F \leq \min \leq \max \leq G$ and, jointly with N, also verify Kleene's law, $F(x, N(x)) \leq G(y, N(y))$, for all x, y in [0, 1]. Reciprocally, provided F, G, and N were to satisfy the former conditions, the functionally expressed operations defined by $\mu \cdot \sigma = F \, o \, (\mu \, x \, \sigma)$, $\mu + \sigma = G \, o \, (\mu \, x \, \sigma)$, and $\mu' = N \, o \, \mu$, would endow $[0, 1]^X$ with a BAF structure. Thus, the former properties are necessary and sufficient for counting with a decomposable BAF, and, because there is a big multiplicity of triplets (F, G, N) verifying such properties, the number of BAFs that can be defined on $[0, 1]^X$ is just enormous and, at each practical problem requiring decomposable "and", "or", and "not", additional properties should be added to functions F, G, and N, for the goal of counting with a calculus. If such additional properties should be imposed according to what can be checked in language, for not supposing properties nonexistent among its words and statements, it is worth noticing that in some practical cases it could be unnecessary to take the three operations decomposable, but just some of them. Let's repeat anew that such decisions are but a matter of design in each particular problem or subject.

(k) Concerning the distributive laws, $\mu \cdot (\alpha + \beta) = \mu \cdot \alpha + \mu \cdot \beta$, and $\mu + (\alpha \cdot \beta) = (\mu + \alpha) \cdot (\mu + \beta)$, it is obvious that they hold simultaneously when $\cdot$ = min and + = max. Nevertheless, the first holds for all μ, α, and β; taking $\alpha = \beta = \mu_1$, it implies $\mu + \mu = \mu$, for all μ, and hence + = max, regardless of concerning the second, $\alpha = \beta = \mu_0$ implies $\mu \cdot \mu = \mu$, for all μ, thus $\cdot$ = min regardless of +.
Hence, with + = max, the distributive law of $\cdot$ to + holds, and with $\cdot$ = min that of + to $\cdot$; of course, both only hold simultaneously with min and max. If both laws simultaneously hold in arithmetic, in plain language its joint validity cannot always be supposed.

4.5. To end this chapter, let's do a last reflection on what corresponds to assigning an algebraic structure to the membership functions. For such a goal, it can be illustrative enough to look at the case with the triplet (min, max, 1 − id), the most frequently used in the applications of fuzzy sets, and in which most of the Boolean laws are preserved, as they are, for instance the two distributive laws, but not those of noncontradiction $\mu \cdot \mu' = \mu_0$, and excluded-middle $\mu + \mu' = \mu_1$, the only Boolean laws that do not hold with the connectives min and max. The distributive laws were already rejected in the reasoning physicists conduct on quantum mechanics with their specialized theoretical language, based on the algebra of a Hilbert space and, although holding with precise words, a general reason is not even known for its validity when imprecise words are used in language; hence its supposition can violate what is behind a description of something in plain language. By just incorrectly supposing one of them, there is added to language a law that can lead to conclude something not real, and which can introduce a lack of confidence on the corresponding conclusions. That distributive laws don't hold in the language of quantum physics suffices to show that these laws cannot always be taken for granted.

Analogous comments can be made for any law not appearing in the list of axioms defining a BAF, therefore each time one such addition seems to be necessary for computing, it should be carefully checked in each case that can imply its addition. Note that, in fact, the laws presumed in a BAF are almost coincidental with those that were needed to reach some properties of meaning.

All this shows the importance of paying careful attention to the design of the membership functions and the connectives that could appear in such design, by previously acquiring the best available information on the contextual behavior of their linguistic labels; that is, to a good comprehension of what they actually express. There is no universal algebra with imprecise words.

References

1. C. Alsina, E. Trillas, L. Valverde, On some logical connectives for fuzzy sets theory. J. Math. Anal. Appl. **93**, 15–26 (1983)
2. E. Trillas, C. Moraga, S. Termini, A naïve way of looking at fuzzy sets. Fuzzy Sets Syst. **292**, 380–393 (2016)
3. M.H. Stone, The theory of representation of Boolean algebras. Trans. Amer. Math. Sci. **40**, 37–111 (1936)
4. P.R. Halmos, *Naive Set Theory* (Van Nostrand, New York, 1960)
5. E. Trillas, A model for 'crisp' reasoning with fuzzy sets. Int. J. Intell. Syst. **27**(10), 859–872 (2012)
6. E. Trillas, On a Model for the Meaning of Predicates, in *Views on Fuzzy Sets and Systems from Different Perspectives*, ed. by R. Seising (Springer, Berlin, 2009), pp. 175–205
7. L.A. Zadeh, Fuzzy sets. Inf. Control **8**, 338–353 (1965)
8. E. Trillas, S. Cubillo, J.L. Castro, Conjunctions and disjunctions in [0, 1]. Fuzzy Sets Syst. **72** (2), 155–165 (1995)
9. S. Guadarrama, E. Renedo, E. Trillas, A first inquiry on semantic-based models of *And*, in *Accuracy and Fuzziness*, ed. by R. Seising, L. Argüelles (Springer, Heidelberg, 2015), pp. 331–350
10. H.T. Nguyen, E.A. Walker, *A First Course in Fuzzy Logic* (Chapman and Hall, Boca Raton, 2000)
11. G.J. Klir, B. Yuan, *Fuzzy Sets and Fuzzy Logic. Theory and Applications* (Prentice-Hall, Upper Saddle River, 1995)
12. G. Böhme, *Fuzzy-Logik* (Springer, Berlin, 1993)
13. H.J. Zimmermann, P. Zysno, Latent connectives in human decision making. Fuzzy Sets Syst. **4**, 17–51 (1980)
14. S. Weber, A general concept of fuzzy connectives, negation, and implication based on t-norms and t-conorms. Fuzzy Sets Syst. **11**, 115–134 (1983)
15. E. Trillas, L. Eciolaza, *Fuzzy Logic* (Springer, Berlin, 2015)
16. C. Alsina, E. Trillas, On the symmetric difference of fuzzy sets. Fuzzy Sets Syst. **153**(2), 181–194 (2005)
17. E. Renedo, E. Trillas, C. Alsina, A note on the symmetric difference in lattices. Math. Soft Comput. **12**, 75–81 (2005)

18. E. Renedo, E. Trillas, C. Alsina, On the law $(a \cdot b')' = b + a' \cdot b'$ in De Morgan and orthomodular lattices. Soft. Comput. **8**(1), 71–73 (2003)
19. I. García-Honrado, E. Trillas, Remarks on the symmetric difference from an inferential point of view. Multiple-Valued Logic Soft Comput. **24**(1), 35–57 (2014)

Chapter 5
A First Look at Conditional Statements

Most, if not all knowledge is constrained by some former knowledge, for instance, as with common expressions such as "John dresses a nice jacket," whose understanding is constrained by previously capturing what "dress", "jacket", and "nice" mean in plain language and in the context of such an utterance, or "All prime numbers are odd," the understanding of which needs to know previously what is understood by "numbers", "prime", and "odd" in the language of arithmetic.

Most of these statements can be expressed in the form "*If* this, *then* that"; for instance, "If (divisors of a prime number are only itself and one, and an odd number is one whose division by two gives a remainder of one), then (all prime numbers are odd)." They are not simply assertive statements, but conditional, or, if/then-statements, and their good comprehension shows some evidence of maturity; in childhood, for instance, it often takes some time to understand fully if/then, hypothetical, statements. Analogously, most questions are of a conditional type, such as "Because today is very cloudy and wet, will there be a flood?" or "If the connectives are prod and W, do the distributive laws hold?"

Hence the analysis of conditional statements' meaning is paramount for studying reasoning under a symbolical representation. Knowledge always implies some former knowledge, even for establishing the Peano axioms defining the positive integers, or those defining an ortholattice. Obviously, sometimes the previous knowledge can be of a very different nature of the sequent one; in physical subjects, for instance, the former knowledge can be of an experiential type and the sequent one of a purely formal one, and in mathematical subjects the previous knowledge can be of a formal type.

Actually, and at least partially, new knowledge comes from some older knowledge even if, in ordinary reasoning, it is not always possible to reduce it to a few axioms such as in the case of arithmetic with those of Peano, in which all possible new knowledge is already compacted and deductively extracted from it, although sometimes with the help of a sophisticated mathematical methodology. It should be noticed that the "artificial" language of arithmetic is not only translatable

E. Trillas, *On the Logos: A Naïve View on Ordinary Reasoning and Fuzzy Logic*,
Studies in Fuzziness and Soft Computing 354, DOI 10.1007/978-3-319-56053-3_5

into plain language, but also basically thought in it. Plain language is indeed the way we have for doing almost everything.

5.1. Let $(X, <_P, m_P)$ and $(Y, <_Q, m_Q)$ be the respective meanings of P in X and Q in Y, and suppose that the relation

$$R(P,Q) \subseteq X[P] \times Y[Q]$$

represents a linguistically expressed relationship between the statements "x is P" and "y is Q"; for instance, "No prime number different from 2, is even," or "With too much sugar, beverages are not healthy." Then, we can consider the "constrained", or "relational", predicate Q/P (Q if P, or Q provided P) naming the relation $R(P, Q)$, by defining

$$(x,y) \text{ is } Q/P \Leftrightarrow (x \text{ is } P, y \text{ is } Q) \in R(P,Q).$$

Relational predicates Q/P are essential for almost everything and, sometimes, they receive a specific name. In the logical calculus, for instance, Q/P is understood as "If P, then Q" and in the Boolean case it is usually identified with "not P or Q", that is, $Q/P = P' + Q$. In plain language there are cases in which Q/P is understood as "P and Q"; that is, Q/P is identified with $P \cdot Q$, and so on.

There are also cases in which Q/P is simply indicative of a possible action and sometimes understood as a command; for instance, "There is an ashtray close to the hand [in which] you hold the cigarette." Anyway, and usually, it is not Q/P coincidental with P/Q.

In general, the qualitative meaning of Q/P in $X \times Y$, $<_{Q/P}$, should be linked with the respective meanings $<_P$, and $<_Q$, in forms such as it verifies

$$<_{Q/P} \subseteq <_P x <_Q,$$

an inclusion not guaranteeing the measurability of Q/P because this inclusion cannot avoid the possibility $<_{Q/P} = \emptyset$. Anyway, provided $<_{Q/P}$ were not empty, there would be some possible definitions, including

$$<_{Q/P} = <_P \times <_Q, \quad \text{or} \quad <_P^{-1} \times <_Q, \quad \text{or} \quad <_P \times <_Q^{-1}, \quad \text{or}$$
$$<_{Q/P} = <_P^{-1} \times <_Q^{-1},$$

and the like, each facilitating obtaining possible measures $m_{Q/P}$ in the function of the corresponding measures m_P and m_Q. For instance, in the first case, and provided the measure were decomposable (i.e., there would be a numerical function J: $[0, 1] \times [0, 1] \to [0, 1]$, such that $m_{Q/P} = J \circ (m_P \times m_Q)$), it suffices a nondecreasing J in its two variables to have:

$$(x_1, y_1) <_{Q/P} (x_2, y_2) \Leftrightarrow x_1 <_P x_2 \;\&\; y_1 <_Q y_2 \Rightarrow m_P(x_1) \leq m_P(x_2) \;\&\; m_Q(y_1) \leq m_Q(y_2)$$
$$\Rightarrow J(m_P(x_1), m_Q(y_1)) \leq J(m_P(x_2), m_Q(y_2)) \Leftrightarrow m_{Q/P}(x_1, y_1) \leq m_{Q/P}(x_2, y_2).$$

An analogous result can be obtained in the second case, with a function J being decreasing in its first variable, and nondecreasing in the second. Of course, to state that the functions obtained in this way are actually measures of Q/P, the relationship between maximal and minimal elements is lacking and thus unknown. For instance, and in the second case, provided that a maximal (x, y) for $<_{Q/P}$ were to imply that x is a minimal for $<_P$, and y a maximal for $<_Q$, then it would suffice taking a function J decreasing in its first variable, nondecreasing in the second, and verifying $J(0, 1) = 1$, because then $m_{Q/P}(x, y) = J(0, 1) = 1$, and so on are with it.

As always, it is clear that the measures of the constrained predicates should be obtained by a careful analysis of the behavior of their qualitative meaning, and its relation to those of P and Q; no general recipe exists for it, and their decomposability is but a hypothesis to be checked, in each case, within the underlying context.

5.2. Several linguistic interpretations can be done to Q/P as it is, for instance, the before-mentioned "not P or $Q = P' + Q$"' that, even not being the unique existing interpretation, deserves a stop because it is usually considered as such in the calculus with precise words. In the quantum calculus Q/P is interpreted by "not P or (P and Q)" $= P' + P \cdot Q$ that, in the classical case in which distributive laws hold, is just reduced to $(P' + P) \cdot (P' + Q) = P' + Q$, because the classical excluded-middle principle implies $(P + P') \cdot S = S$ for all S. This leads to the observation that, as with the linguistic connectives "and", "or", and "not", there is no universal form of expressing if/then statements in plain language; for instance, and as said before, there are contexts in which they are understood by "P and Q". Note that were the conjunction "and" commutative, it would follow the oddity $Q/P = P/Q$.

The undisputed algebraic structure allowing us to model what is relevant for computing with precise words, is that of the calculus in Boolean algebras. In them, many formulae deduced from their axioms represent the patterns of the exact reasoning with such words, and it deserves some comments concerning the representation of conditional statements.

Once a constrained predicate Q/P is represented in a Boolean algebra by the operation $p \to q = p' + q$ (p representing "x is P", and q representing "y is Q"), and the affirmation of "(x, y) is Q/P", is understood as $p' + q = 1$ (if p, then q, is a tautology), it is easy to deduce that this is equivalent to $p \leq q$, the natural order relation of the algebra, coming from its lattice's part, and defined by $p \cdot q = p$ or, equivalently, by $p + q = q$. In fact, if $p' + q = 1$, it follows that $p \cdot (p' + q) = p \cdot q = p$; and if $p \leq q$, it follows that $q' \leq p'$, and then $q' + q = 1 \leq p' + q$; that is, $p' + q = 1$. With this sequence of steps, the relation $R(P, Q)$ is simply reduced to be represented by the relation of partial order $\leq$, independently of which the words P and Q can be. With it, the theoretical desiderata, typical of mathematics, to count with a universal and not semantically dependent symbolism, is reached at the cost of only attending to syntax.

Anyway, and for what corresponds to proving mathematical theorems, the model seems to be pretty good even if some mathematicians did claim against some of its suppositions, the so-called "intuitionists," who, not accepting the strong character of negation, place themselves in a setting weaker than a Boolean algebra, and, consequently, refuse the abundant proofs obtained by reduction to absurdity, and only accept those that, at least potentially, can show how a solution could be effectively constructed step by step. Nevertheless, almost all working mathematicians don't adhere to intuitionism's claims, and remain anchored in classical Boolean methodology.

The situation is actually different concerning the search prior to a proof, in which case mathematicians proceed through the (apparently) disorganized common sense reasoning as do ordinary people. On the other hand, if computing machines prove their capability of algorithmically proving theorems, the machines still show a serious lack of imagining or guessing possible solutions for a given problem. Guessing is still a challenge for computers, and without overcoming it, computers will not be able to reason like people do and the Gordian knot of artificial intelligence will remain unopened.

Concerning the so-called logic of quantum physics represented in orthomodular lattices, lacking the Boolean distributive laws, constrained statements are often represented by the operation $p' + p \cdot q$, the Sasaki hook, that once made equal to 1 (the lattice maximum) also becomes equivalent to $p \leq q$, and a desiderata such as the former is reached. Another form used for representing a conditional is the Dishkant hook $q + (p' \cdot q')$, also leading to the order of the lattice.

Nevertheless, in the case of De Morgan algebras, neither $p' + q$, nor $p' + p \cdot q$ are useful for representing constrained statements Q/P by such operations $p \to q$, due to not permitting the modus ponens scheme to hold, and usually translated into the inequality $p \cdot (p \to q) \leq q$, allowing us to state that $p = 1$, and $p \to q = 1$, gives $q = 1$ (the assertions of P and Q/P assure that of Q). In fact, provided it were $p \cdot (p' + q) \leq q$ for all p and q in the De Morgan algebra, by just taking $q = 0$ it would follow that $p \cdot p' = 0$ for all p, implying the absurdity that the De Morgan algebra is just a Boolean one, and it analogously happens with $p' + p \cdot q$, because $q = 0$ also implies $p \cdot p' = 0$. It is for this reason that different expressions should be used in De Morgan algebras for representing conditionals.

One such representation comes from the fact that, in complete Boolean algebras, it is $p' + q = \text{Sup}\ \{z; p \cdot z \leq q\}$, as proven by the following steps.

(a) $p \cdot (p' + q) = p \cdot q \leq q$, proving that $p' + q$ is among the z.
(b) Provided a bigger z were to exist, that is, such that $p' + q \leq z$, and $p \cdot z \leq q$, it would follow that $p' + p \cdot z = p' + z \leq p' + q$, and, because it is $z \leq p' + z$, it follows that $z \leq p' + q$, and hence it is proven that $z = p' + q$.

Then, a possibility consists in taking, in complete De Morgan algebras, $p \to q = \text{Sup}\ \{z; p \cdot z \leq q\}$, after checking that it verifies the MP inequality.

Notice that it is obvious provided $p \to q$ were attainable, that is, if the supremum were just the maximum of the set $\{z;\, p \cdot z \le q\}$.

5.3. What does *Q/P* represent when *P* and *Q* are, respectively, just known by their membership functions, but no universal representation of the conditional can be supposed? The usual way consists in a first step accepting the hypothesis that the membership function of *Q/P*, $\mu_{Q/P}$, is decomposable, after doing some checking of that aspect. The second step consists in finding a suitable numerical function *J* of two arguments, giving

$$\mu_{Q/P}(x, y) = J\big(\mu_P(x), \mu_Q(y)\big), \quad \text{for all } x \text{ in } X, \quad \text{and } y \text{ in } Y;$$

"suitable" refers here to being in accordance with the contextual linguistic interpretation of *Q/P*. Finally, the third step consists in checking that the constructed function verifies the inequality of modus ponens for some numerical function *F*:

$$F(\mu_P(x), J\big(\mu_P(x), \mu_Q(y)\big) \le \mu_Q(y), \quad \text{for all } x \text{ in } X, \quad \text{and } y \text{ in } Y,$$

for it suffices to study if the numerical inequality

$$F(a, J(a, b)) \le b \ \text{ holds for all } a, b \text{ in } [0, 1],$$

by searching for the functions F allowing its validity.

For instance, provided *Q/P* could be interpreted as "Not *P* or *Q*", the function *J* should be of the type $J(a, b) = G(N(a), b)$, with selected suitable functions *G*, representing "or", and *N*, representing "not", and, once both are chosen, a modus ponens function *F* for the verification of $F(a, G(N(a), b)) \le b$, should also be found. Provided the representation of *Q/P* were to correspond to "Not *P*, or *P* and *Q*", it should be $J(a, b) = G(N(a), F(a, b))$, and a suitable modus ponens function F^* verifying $F^*(a, G(N(a), F(a, b)) \le b$, for all *a*, *b* in [0, 1] should be also found.

Provided none of these models, or proto-forms, were adequate, there would still be the possibility of reverting to a function $J_F(a, b) = \text{Sup}\ \{z \text{ in } [0, 1];\ F(a, z)\ \ b\}$; such a case happens when, for instance, it is very difficult to describe linguistically not $P = P'$, making the design of a membership function with such a linguistic label extremely difficult.

5.4. The representation of if/then statements is certainly relevant in the applications of fuzzy sets to control systems inasmuch as there are many dynamical systems that can be linguistically described by means of imprecise rules that, once established by an expert and translated into fuzzy terms, serve for computer simulation and also for the practical reproduction of the actual system's behavior, the basis of industrially successful fuzzy control.

It is with such linguistic rules, "If antecedent, then consequent," very often expressed in plain language with imprecise words, that the design of the negation of their antecedents is sometimes very difficult, if not impossible, by referring to several possible and different situations. Hence, they should be either represented by a J_F function, or by another function J also not containing the antecedent's negation.

The most used such functions in fuzzy control are $J(a, b) = \min(a, b)$, and $J(a, b) = a. b$, that, coming from interpreting "If P, then Q" as "P and Q", generate the methods called, respectively, Mamdani and Larsen. They are, notwithstanding, commutative operations, and consequently, force the equivalence between Q/P and P/Q, contrary to the meaning of rules that are not usually reversible.

These functions verify the MP inequality with $F = \min$ (e.g., it is $\min(a, a. b) \leq \min(a, b) \leq b$) and there is an easy form for transforming these operations into another, but not commutative, J:

$$J(a,b) = \min(a^{\lambda}, b), \quad \text{and} \quad J(a,b) = a^{\lambda}.b, \quad \text{with } \lambda > 0.$$

Both satisfy the MP inequality with $F = \min$: for instance, $\min(a, \min(a^{\lambda}, b)) \leq \min(a^{\lambda}, b) \leq b$. It corresponds to interpreting Q/P in a "conjunctive" form, $Q/P = P$ and Q, with a noncommutative "and", as is typical of plain language; that is, it corresponds to a linguistic interpretation of if/then statements proper to plain language and reached by modifying the antecedent of the rules by taking $\mu_P(x)^{\lambda}$ instead of $\mu_P(x)$ at each point x.

A general and formal interpretation of Q/P is done by the general operation with connectives $p \to q = p' \cdot q + p' \cdot q' + p \cdot q$, a proto-form that, in a Boolean algebra is reduced as follows.

$$p \to q = p' \cdot (q + q') + p \cdot q = p' + p \cdot q = (p' + p) \cdot (p' + q) = p' + q.$$

Note that the linguistic expression of the former first equation is

"It is not p and it is (q or not q), or it is (p and q)",

covering all the possibilities for stating Q/P with connectives. For instance, the conditional statement "If it is raining, I take my umbrella," means, "It is not raining and (either I take my umbrella or not), or (it is raining and I take my umbrella)" and so on.

A still more general operation is

$$p \to q = \alpha \cdot p' \cdot q + \beta \cdot p' \cdot q' + \gamma \cdot p \cdot q,$$

with α, β, and γ in the Boolean algebra, and of which the former proto-forms are but particular cases: for instance, with $\alpha = \beta = \gamma = 1$, $p' + q$ is obtained, and with $\alpha = \beta = 0$, $\gamma = 1$, $p \cdot q$ is obtained. Note that this (conjunctive) proto-form $p \cdot q$ comes from avoiding the term $p' \cdot q + p' \cdot q' = p'$.

5.5. When F is a continuous t-norm, it is $J_F(a, b) = 1 \Leftrightarrow a \leq b$, but it is neither that the J of Mamdani, nor that of Larsen, verifies such equivalence. In fact, min $(a, b) = 1$ is just equivalent to $a = b = 1$, and the same happens with prod $(a, b) = 1$. It can be said that functions J_T order [0, 1], but that neither min nor prod order it. It analogously happens with the functions of the types $G(N(a), b)$, and $G(N(a), F(a, b))$; for example, $\max(1 - a, b) = 1 \Leftrightarrow a = 0$ or $b = 1$, and $\max(1 - a, \min(a, b)) = 1 \Leftrightarrow a = 0$ or $a = b = 1$.

Of course, these notions can be translated into $[0, 1]^X$, and in this sense, it is $J_F(\mu(x), \sigma(x)) = 1$ for all x in $X \Leftrightarrow \mu(x) \leq \sigma(x)$ for all x in $X \Leftrightarrow \mu \leq \sigma$. Of course, from a mathematical point of view functions J ordering $[0, 1]^X$ can be more exciting than those not ordering it, but things are what they are, and what should be modeled by means of mathematics is the very complex plain language that cannot be flattened by imposing external properties on it, unless they can be internally checked and are actually found to hold. The risk lies in obtaining mathematical models not modeling what is tried to be modeled.

5.6. Let's make a last and short comment on the philosophically troublesome subject of counterfactual conditionals, that is, those whose antecedent is but imaginary. They are typical when joking; for instance, "If this year is 2060, now it is raining," "If the moon is made of cheese, then I am reading a book," "If this book is on the history of Rome, I will learn how Neanderthals disappeared," and so on. Such statements are, actually, meaningless even if some of them can be true under some interpretation, as is the second one because (The moon is made of cheese) + (I am reading a book) = The moon is not made of cheese, or I am writing a book, is currently true. The first, notwithstanding, cannot be currently stated as true, because its antecedent is false and its consequent's truth cannot be fixed inasmuch as it depends on the moment. The third cannot be stated as true because no actual relationship can be presumed between the Romans and the Neanderthals that disappeared before the beginning of Rome's history. Note that interpreting the second conditional in conjunctive form, "The moon is made of cheese and I am reading a book" has no chance of being true.

Notwithstanding, some special counterfactual conditionals could facilitate nondeductive creative reasoning, something that, although it is further considered later, for right now an illustrative example suffices. These counterfactuals appear when no more hypotheses are known to explain something well. The hypothesis is the conditional's antecedent, and its consequent is what is tried to be checked and that, if it actually occurs, will produce more confidence in the hypothesis. It happened, in the old geocentric conception of the heavens, by presuming (the hypothesis) that the planets were doing epicycles from which their positions were calculated and observed by the naked eye. Because such observations by the naked eye confirmed the computed positions very well, it was believed that the conditional "If epicycles, then positions" was correct with its consequent either correct or not correct. Hence, the conditional was interpreted as "Not epicycles, or positions" $(e \to p = e' + p)$, and taking the truth values $t(e \to p) = 1$, and $t(p) = 1$, from $t(e \to p) = \max(1 - t(e), t(p))$, it follows that $1 = \max(1 - t(e), 1)$, holding for any value of $t(e)$; that is, no conclusion on the truth or falsity of the hypothesis can be concluded by presuming the conditional is true. By presuming it is false, the resulting equation $0 = \max(1 - t(e), 1)$ shows, nevertheless, the absurdity $0 = 1$ that leads to perplexity.

Notice that presuming $e \rightarrow p = e \cdot p$, then $t(e \rightarrow p) = \min(t(e), t(p))$ implies $1 = \min(t(e), 1) = t(e)$, that the antecedent is true; it seems that, at least, the "belief" in the truth of the antecedent after presuming that of the conditional and once that of the consequent is checked, is better taken into account by the conjunctive interpretation of the conditional.

This is the kind of troublesome worry counterfactuals produced historically by playing with them in deductive reasoning.

In sum, counterfactuals cannot always be taken into account for safe deduction; they can be, perhaps, considered as a kind of help for guessing. For instance, in old times, it was common to say, "This happens as if it were such and such," equivalent to asserting the conditional, "If such and such, then this," but without asserting the reality of the antecedent "such and such", and only taking it as a working hypothesis allowing computations. This was how Galileo (supposedly) tried to defend himself from theologians, by thinking the famous, "*Eppur si muove*" (Italian: "And yet it moves").

References

1. E. Trillas, I. Garcia-Honrado, A reflection on fuzzy conditionals, in *Combining Experimentation and Theory*, eds. by E. Trillas et al. (Springer, Berlin, 2012), pp. 391–406
2. E. Trillas, C. Alsina, A. Pradera, On mpt-implications for fuzzy logic. Revista Real Academia de Ciencias, Serie A **98**(1), 259–271 (2004)
3. E. Trillas, L. Eciolaza, *Fuzzy Logic* (Springer, Berlin, 2015)
4. E. Trillas, S. Cubillo, Modus ponens in Boolean algebras, revisited. Mathware Soft Comput. **3** (1–2), 105–112 (1996)
5. E. Trillas, C. Alsina, From Leibniz's shinning theorem to the synthesis of rules trough Mamdani-Larsen conditionals, in *Combining Experimentation and Theory,* eds. E. Trillas et al. (Springer, Berlin, 2012), pp. 247–256
6. E. Trillas, On a model for the meaning of predicates, in *Views on Fuzzy Sets and Systems from Different Perspectives*, ed. by R. Seising (Springer, Berlin, 2009), pp. 175–205
7. N. Goodman, The problem of counterfactuals. J. Philosophy **44**(5), 113–128 (1947)
8. E. Trillas, L. Valverde, On implication and indistinguishability in the setting of fuzzy logic, in *Managing Decision Systems by Fuzzy Sets and Possibility Theory,* eds. by J. Kacpryk et al. (TUV-Rheinland, Köln, 1985), pp. 198–212
9. J. Ferrater Mora, H. Leblanc, *Lógica matemática* (Fondo de Cultura Económica, México, 1973)
10. J.G. Hüllsmann, Facts and counterfactuals in economic law. J. Libertarian Stud. **17**(1), 57–102 (2003)

11. E. Trillas, R. Seising, On material and strict implication, in *Archives for the Philosophy and History of Soft Computing*, vol. 2 (2014), pp. 1–12
12. J. Del Val, A. Rivière, Si llueve, Elisa lleva sombrero: Una investigación psicológica sobre la tabla de verdad del condicional. Revista de Psicología General y Aplicada **30**, 825–850 (1975)
13. S. Guadarrama, E. Renedo, E. Trillas, A first inquiry on semantic-based models of *And*, in *Accuracy and Fuzziness*, eds. R. Seising et al. (Springer, Heidelberg, 2015), pp. 331–350
14. R. Stalnaker, A theory of conditionals, in *Studies in Logical Theory*, ed. by N. Rescher (Blackwell, Oxford, 1968), pp. 98–112
15. G. Pólya, *Mathematics and Plausible Reasoning*, vol. I (Princeton University Press, Princeton, 1954)
16. G. Pólya, *Mathematics and Plausible Reasoning*, vol. II (Princeton University Press, Princeton, 1968)
17. E. Trillas, E. Renedo, C. Alsina, On three laws typical of booleanity, in Proceedings of NAFIPS (2004), pp. 520–524

Chapter 6
Linguistic Qualification, Modification, and Synonymy

Affirmative relational statements are often uttered in plain language, such as "The dark tower is spacious," that, within a universe of towers, and with the words P = dark, and τ = spacious, can be represented by "The tower is P, and it is τ," or, shortened by "The tower is P is τ." In such statements, the information P facilitates the elements in the universe X, and is constrained by qualifying by τ. Representing such statements by "x is (P is τ)", it is said that τ linguistically qualifies, or constrains, P in X, and it can be considered that (P is τ) is a new qualified or constrained predicative pair acting on the universe X, and perhaps renamed in it in some shortened but expressive form such as "This tower is dark and spacious."

Qualification is frequent in plain language: "The green sofa is large," "This intelligent person is marvelous," and the like. In all these instances, τ constrains the information P alone facilitates, and it is applied to X [P] in an imprecise, or sometimes precise, form, for instance, "John is tall: he has a height of 190 cm." Hence its abundance in plain language makes the analysis of the simplified statements "x is (P is τ)", for x in X, of some relevance, either if τ is an imprecise, or a precise word, or a complex of words. Qualification is a trick plain language employs for constraining the meaning of imprecise words and facilitating more information. The first question is finding the meaning of "P is τ" in X.

6.1. Let's suppose that P and (P is τ), both applied to X, show the respective meanings $(X, <_P, m_P)$ and $(X, <_{P \text{ is } \tau}, m_{P \text{ is } \tau})$, of which only the first is known. What about the meaning of τ?

It can be supposed that τ is applied to the range of values of m_P simply for the goal of composing its measure with that of P; that is, such meaning is supposed to be $([0, 1], <_\tau, m_\tau)$, and then on which conditions it can be $m_{P \text{ is } \tau} = m_\tau \circ m_P$ is studied. Note that this is a hypothesis corresponding to suppose that τ, in principle applied to X [P] has, as a proper numerical counterpart, a word τ^* that applied to [0, 1], and once specified, is identified with τ. What follows is under this presumption.

E. Trillas, *On the Logos: A Naïve View on Ordinary Reasoning and Fuzzy Logic*,
Studies in Fuzziness and Soft Computing 354, DOI 10.1007/978-3-319-56053-3_6

In the first place, it seems obvious that if it is "x is (P is τ)", it should also be "x is P", with which it is reasonable to suppose only that $<_{P \text{ is } \tau} \subseteq <_P$, an inclusion not assuring the nonemptiness of the first relation, but that provided it were not empty means

$$x <_{P \text{ is } \tau} y \Rightarrow x <_P y \Rightarrow m_P(x) \leq m_P(y).$$

It (newly!) suffices the hypothesis that m_τ: [0, 1] $\to$ [0, 1] is nondecreasing for arriving at $m_\tau(m_P(x)) \leq m_\tau(m_P(y))$, and it only lacks knowing what happens with the maximal and minimal elements with respect to $<_{P \text{ is } \tau}$. But it does not seem bizarre to suppose that all its maximal and minimal elements should also be maximal and minimal, respectively, for $<_P$. Consequently, it suffices to presume $m_\tau(1) = 1$ and $m_\tau(0) = 0$, and then, if x is maximal, $m_{P \text{ is } \tau}(x) = m_\tau(1) = 1$, and if y is minimal, $m_{P \text{ is } \tau}(y) = m_\tau(0) = 0$ and under all these conditions $m_\tau \circ m_P$ is a measure for (P is τ).

How, in the toy example given by P = small in [0, 10] and τ = large in [0, 1], can (small is large) be named? Supposing $m_{\text{small}}(x) = 1 - x/10$ and large identified with the linguistic label of the subinterval (5, 10] in [0, 10], it follows that $m_{P \text{ is } \tau}(x) = m_\tau(1 - x/10)$, and because $m_\tau(x) = 0$, if $x \in [0, 5]$, and 1 in (5, 10], is $m_{P \text{ is } \tau}(x) = 1$, if $x \in [0, 5)$ and 0 otherwise. Thus, under such specification of τ, "x is small is large" $\Leftrightarrow$ "x is less than five" and "small is large" can be called "less than five". Of course, and provided τ were specified by the identity $m_\tau(x) = x$, the meaning of "small is large" would just coincide with that of "small"; under this specification of τ, the qualification does not add a single bit of information to P and τ results in a redundant word.

Note that each time τ is precise, or crisp, (P is τ) also has a precise meaning. To capture the full meaning of (P is τ) well, it is necessary that those of P and τ have previously been captured; provided the meaning of τ in X [P] were not identifiable with that of some τ^* in [0, 1], then the meaning of (P is τ) in X should be directly studied, but without the possibility of taking $m_\tau \circ m_P$ as its measure.

6.2. Let m be an adverb immediately applicable to words P, giving a new and complex word m_P; for instance, if P = tall and m = very, it is m_P = very-tall, or if m = more or less, it is m_P = more or less tall. It is another way of qualifying P by constraining its meaning and serving more information. It is also frequent in plain language, and it is said that m is a linguistic modifier, or a semantic hedge. Provided the meaning of P in X were known, which meaning can be attributed to m_P in X?

As in Sect. 6.1 it is supposed that m acts in [0, 1]; that it is $\leq \subseteq <_m$, and $<_{mP} \subseteq <_{P}$. With it, follows

$$x <_{mP} y \Rightarrow x <_P y \Rightarrow m_P(x) \leq m_P(y),$$

and by supposing anew that m_m is a nondecreasing function [0, 1] $\to$ [0, 1], is

$$m_m(m_P(x)) \leq m_m(m_P(y)).$$

Thus, provided it could be stated that the maximal and minimal elements for $<_{mP}$ are maximal and minimal, respectively, for $<_P$, it would be concluded that $m_m(m_P(x)) = m_n(1) = 1$, if x is a maximal for $<_{mP}$, and $m_m(m_P(x)) = m_m(0) = 0$, if x is a minimal, and provided it were $m_m(1) = 1$, $m_m(0) = 0$. Hence it suffices to take a nondecreasing function m_m with these border conditions, for having that $m_{mP} = m_m \circ m_P$ is a measure for the qualitative meaning of m_P in X.

Inasmuch as functions in [0, 1] only can be below, above, or crossing the identity function $\mathrm{Id}(x) = x$, the two cases of modifiers such that either $\mathrm{id} \leq m_m$, or $m_m \leq \mathrm{id}$, respectively called "extensive" and "contractive" modifiers, are of interest. Among the contractive are those with $m_m(x) = x^2$, and among the extensive those with $m_m(x) = +\sqrt{x}$, that were considered by Lotfi A. Zadeh when he, for the first time, introduced the possible membership functions of "very P" and "more or less P", respectively. Note that, for instance, $\mathrm{id} \leq m_m$ implies $m_P \leq m_m \circ m_P = m_{mP}$, and $m_m \leq \mathrm{id}$ implies $m_{mP} \leq m_P$, inequalities justifying the names given to the corresponding modifiers.

Of course, instead of these two functions, other functions being either contractive or expansive can be obviously taken to modify P accordingly with the contextual use of the corresponding words, for instance, instead of x^2, the function equal to 0 up to 1/3, followed by a straight line segment from (1/3, 0) to (2/3, 1), and continuing with value 1 up to the border 1 of the unit interval.

Let's show, for instance, a toy example with m = very and P = short in [0, 10]. It can be taken $m_{\text{very}}(x) = x^2$, because "very" seems to be a contractive modifier, and thus $m_{\text{very short}}(x) = (1 - x/10)^2 = x^2/100 - 2x/10 + 1$. Because "more or less" seems to be an expansive modifier, it can be taken $m_{\text{more or less}}(x) = +\sqrt{x}$, and then it is $m_{\text{more or less small}}(x) = +\sqrt{(1 - x/10)}$ and so on.

6.3. Synonymy is a semantic, contextual, and important linguistic phenomenon in plain language that, almost nonexistent in the artificial languages, refers to closeness of meaning. Two words P and Q are said to be synonyms if their meanings are almost coincidental; obviously, for a deep study of synonymy what "almost coincidental" could mean should be tried to be clarified, and how it can be represented in mathematical terms.

It should be pointed out that, as can be easily seen by looking at a dictionary of synonyms, the chains of synonyms break in a short number of steps; if word P is synonymous with Q, and Q with R, and so on, it always arrives at a word Z that is not a synonym of P. That is, the semantic relation of synonymy is not transitive and, in some cases, it is not even symmetrical because the context could be such that P can be taken as a synonym of Q, but Q cannot be taken as a synonym of P. Meaning is, along a chain of synonyms, sequentially flattened in such a way that ends by losing any meaning's similarity with the first word. This seems to note that the semantic phenomenon of synonymy is essentially linked with imprecision, and that with precise words the only existing synonymy is due to a large intersection of the sets representing the words, that is, an almost coincidence of meaning and also something of an imprecise character unless there is coincidence of both sets.

Synonymy is reconsidered later on and from a different point of view. But, because synonymy is actually related to meaning, it is of some interest to advance a few previous remarks on it. When a pair of words P and Q can be seen as synonyms, it is also said (perhaps abusively) that "P means Q", or that "Q means P". Let's recall that, in principle, synonymy is not always a symmetrical relation.

Let P be with $(X, <_P, m_P)$, and Q with $(Y, <_Q, m_Q)$. It is said that Q is an f-synonym of P if a function $f: X \to Y$ exists and verifies:

$$x_1 <_P x_2 \Leftrightarrow f(x_1) <_Q f(x_2).$$

With it and provided f were bijective (one to one and onto), because then it would be, equivalently, $y_1 <_Q y_2 \Leftrightarrow f^{-1}(y_1) <_P f^{-1}(y_2)$, the respective qualitative meanings would be equivalent. It is

$$<_Q = <_P \circ (f^{-1} \times f^{-1}),$$

and their respective measures can be linked by

$$m_Q = m_P \circ f^{-1}; \text{ that is, } m_Q(y) = m_P(f^{-1}(y)), \text{ for all } y \text{ in } Y,$$
$$\text{or } m_P(x) = m_Q(f(x)), \text{ for all } x \text{ in } X.$$

Note that if f is not bijective but just one to one, only this last equality can be considered. When $X = Y$, and $f = id_X$, a pair of id-synonyms is of exact synonyms; two different words endowed with the same meaning.

For instance, the toy example given by P = small in [0, 1], and Q = short in [0, 10], shows that $m_{short}(y) = m_{small}(x/10) = 1 - x/10$ is with $f(x) = 10x$; hence, small and short can be considered f-synonyms with $f = 10 \times \text{Id}$.

The f-synonymy between P and Q could be symbolically written $(Y, <_Q, m_Q) = f(X, <_P, m_P)$, or, if f is bijective, $(X, <_P, m_P) = f^{-1}(Y, <_Q, m_Q)$, to mark the functional relation between the respective meanings of P and Q, and with the equal sign holding if $X = Y$, and $f = id_X$ to reflect the case of exact synonymy.

It should be pointed out that not all pairs of synonyms in plain language are just f-synonyms; the linguistic phenomenon, involving approximate meaning, is more complicated and requires gradation, but for what refers, at least, to the preservation of meaning between two different universes of discourse, as a type of family resemblance between words, f-synonymy is illustrative.

References

1. E. Trillas, On a model for the meaning of predicates, in *Views on Fuzzy Sets and Systems from Different Perspectives*, ed. by R. Seising (Springer, Berlin, 2009), pp. 175–205
2. E. Trillas, C. Moraga, A. Sobrino, On family resemblances with fuzzy sets, in Proceedings of IFSA-EUSFLAT (2009), pp. 306–311

3. G. Lakoff, Hedges: A study on meaning criteria and the logic of fuzzy concepts. J. Philos. Logic **2**(4), 458–508 (1973)
4. L.A. Zadeh, A fuzzy set theoretic interpretation of linguistic hedges. J. Cybernetics **2**(3), 4–34 (1972)
5. M. De Cock, E.E. Kerre, A context-based approach to linguistic hedges. Int. J. Appl. Math. Comp. Sci. **12**(3), 371–382 (2002)
6. S. Pinker, *Words and Rules* (Basic Books, New York, 1999)
7. E.E. Kerre, M. De Cock, Linguistic modifiers. An overview, in *Fuzzy Logic and Soft Computing,* ed. by E.E. Kerre et al. (Springer, New York, 1999), pp. 89–95
8. M. De Cock, E.E. Kerre, Fuzzy modifiers based on fuzzy relations. Inf. Sci. **160**(1), 173–182 (2004)

Chapter 7
Thinking, Analogy, and Reasoning

Thinking is a natural phenomenon generated in the human brain thanks to its neural networks, and whose experimental and theoretic study corresponds to the neurosciences. Speaking and reasoning are but external manifestations of thinking in which both the configuration of the throat and the brain's functioning play decisive roles; if thinking is not always directed by the person, speaking and reasoning are often so. Thinking could be unconscious, but reasoning is almost always conscious and directed to a goal, and for many years attempts have been made to study and mathematically model it; the history of logic contains such an evolution from old philosophy to modern mathematical logic. Reasoning is deeply related to rationality, and at the end, the old Greek term *Logos* referred to both word and reason; we refer to conscious reasoning.

Usually, the natural interlinking of speaking and reasoning is seen as the ground on which rationality is anchored; and it was in the field of psychology where the disturbances, and illnesses, of speaking and reasoning began to be studied. The term "Logos" referred to both people's capability of conducting articulate reasoning, and the use of "the word" to speak for communicating the reasoning and its conclusions. Now, once a new theoretic presentation on how to build up complex statements by capturing their meanings is done, a moment for advancing a view on reasoning seems to have arrived.

7.1. It should be first pointed out that by "reasoning" it is here understood the kind of reasoning laypeople consciously conduct, a reasoning that scarcely shows all the characteristics of the formal and deductive reasoning mathematicians manage for proving their conclusions, the theorems constituting the corpus of mathematics. Ordinary, plain, everyday, or commonsense reasoning (for short, sometimes and afterwards, reasoning), is often not formally deductive, even if it can sometimes present some features approaching formal deduction. Nevertheless, a first characteristic of ordinary reasoning is the "unsafe" character of its conclusions, because (contrary to formal deduction among whose conclusions, or consequences, contradictory pairs of them can never be found) this is not at all rare in the ordinary

E. Trillas, *On the Logos: A Naïve View on Ordinary Reasoning and Fuzzy Logic*,
Studies in Fuzziness and Soft Computing 354, DOI 10.1007/978-3-319-56053-3_7

reasoning done by people. Formal deduction is reasoning with a safety net, but commonsense reasoning is without; it is always like walking in a loose robe.

In part, it was the importance Euclid's *Elements* acquired during the Middle Ages as a model for performing correct reasoning that identified correct reasoning with the safe, and beautiful, deductive reasoning of geometry. Nevertheless, what should essentially be preserved from such a view is just the impossibility of conducting any reasoning without previously accounting for some actual information on the corresponding subject; without such information, for instance, the axioms in geometry, no geometrically interesting conclusions of reasoning are possible. Reasoning cannot be based on an empty set of informative items; to start from some premises reflecting the information, or evidence, is strictly necessary for concluding something actually significant.

The specification of the premises can come from some previous, and perhaps disorganized, thinking on a related question that can involve images, not only words; nevertheless, and in the end, the premises should be specified by words. Without specifying them by words it is difficult, if not impossible, to know if there are, or are not, contradictions among them, that is, if the starting information is liable. Reasoning is done in an environment of knowledge and inside thinking.

The set of premises is just constrained to not contain contradictions, that is, neither pairs of contradictory premises, nor self-contradictory ones, something accepted as a minimal caution for conducting any reasoning. Nobody admits that contradictions among the evidence can be accepted; before starting to reason, it should have been secured that no contradiction exists among its premises. Were a pair of contradictory premises found, at least one of them should be excluded; a previous analysis of the premises' lack of contradictions is compulsory.

The category of liable information is never assigned to a set of contradictory premises on a given subject, and in general to contradictory statements whatsoever. For instance, in a conversation on the state of today's weather, which information could actually be furnished by two statements such as "Right now there are very dense clouds," and "Right now it is totally sunny"? Who will accept these statements as an actual piece of information on the current weather's state? Contradiction is the worst sin of reasoning; it should be prohibited in, at least, the premises, and, provided two contradictory conclusions were obtained at the end of the reasoning, at least one of them should finally be rejected, and the reasoning reviewed.

Two statements p and q are said to be contradictory if the conditional statement holds, "If p, then not q", and q and p are contradictory provided "If q, then not p" holds; p is self-contradictory provided "If p, then not p" holds. Contradiction is not always a symmetric relation even if it can happen under some conditions such as those in Boolean calculus, in which $p \leq q' \Rightarrow (q')' \leq p' \Leftrightarrow q \leq p'$ holds, and reciprocally; nevertheless, it should be pointed out that this proof is made by presuming that stating "If p, then q" is represented by the lattice's order $\leq$ of the Boolean algebra in the form $p \leq q$, and that negation is strong, $(p')' = p$, for all p. Note that it will also hold provided the negation were to be weak, that is, it verifies $q \leq (q')'$, for all q. In a path towards analyzing ordinary reasoning, it seems essential to have a previous reflection on how people can obtain relationships among the meanings of the intervening words for arriving at conceptualizing them and for managing such meanings to reason. In principle, it can be said that reasoning needs a linkage of meanings with something in common, allowing a view of a real or virtual situation; when the "overlapping of meanings" is large enough, the idea of synonymy appears and shows that a reasoning by synonymy will stop. In some way, all reasoning is but a kind of chain of statements whose meanings are related in some form. Only in the deductive/proving reasoning of mathematics are the steps of such chains perfectly linked each one with the following in such a form that not a single hole can exist between them. In common reasoning, the existence of holes, or jumps, is not at all rare, and hence the relationship between its conclusions and its premises is not always clear; only in the case of formal deduction, can the reasoning be perfectly repeated step by step by another person.

If common reasoning could be compared with a house that is badly constructed without the help of an architect, and without taking too much care of its sustaining structure, formal deduction is comparable to a house perfectly well constructed by a competent architect. In the first, removing a single brick can cause the failing of the house, but in the second and for its failing, removing much more than a brick is necessary. Formal deduction requires a perfect enslaving of statements, but in the ordinary forms of reasoning such a kind of chain is often far from being perfect.

People's reasoning is, very rarely, of a totally deductive type, even if people try, in what it is possible, to approach formal deduction; at the end, formal deduction is but a way to control the effectiveness of a reasoning. The uncontestable fact is that formal deduction is the safest type of reasoning, but it requires symbols working in an as "natural" as possible mathematical framework. Hence to attempt a symbolic study of ordinary reasoning allowing its classification in its diverse types and, at the end, clarifying what reasoning actually consists in, it is necessary to know previously how reasoning is initially fueled, and which are its main natural engines.

7.2. Analogy seems to be the first natural engine facilitating reasoning. Analogy is based on the brain's human capability for capturing partially overlapping resemblances among the real, or virtual, objects that are taken into account; it is the linkage of such overlapping meanings that permits reasoning.

Note that thinking is larger than reasoning, because it generates and comprises, for instance, wishing, imagining, memory or storage of information, emotions, dreaming, mixing of old images, and so on, all of them useful for doing reasoning. If thinking is possible without articulate language, conscious reasoning is not, and analogy moves between thinking and ordinary language, inasmuch as without previous conceptualization and naming concepts, it seems very difficult for analogy to support reasoning extensively.

Speaking a language of which reasoning is the most important feature is shared by all people on Earth. From childhood, people simultaneously learn to speak well one, two, or even more languages, to appreciate metaphors, to establish analogies and express them with words, to manage conditionals, and so on. With all this, people learn how to reason, and learn to feel perplexed by an absurd conclusion;

they informally learn how dangerous a contradiction is, and that one should flee from it. The simultaneous learning of both language and reasoning is, in good part, a social phenomenon whose physical possibility lies in that of thinking which all sane children naturally have. All that becomes as it actually appears, due to the brain's physical configuration and functioning that controls the body, allows speaking and thinking, as well as their manifestations such as singing and reasoning, and even enjoying or hating something.

It should be pointed out that one of the main goals of reasoning is satisfactory support for the taking of decisions, and particularly to decide "intelligent" actions; reasoning is basically an important and natural tool with which people are endowed for their survival, not only of each person but, and essentially, of the species *Homo*. This is the reason for considering that the essential characteristic of *Homo*, differentiating its members from those in other animal species, is rationality, the *Logos* externally expressed by telling stories, posing questions, guessing answers, and by computing. Our only way of knowing how to do things, how things are, and how situations will evolve, is reasoning; its possibility is of the greatest help for satisfying human curiosity towards knowing and understanding what is still unknown, or not yet understood, for foreseeing and for answering the questions people pose continuously. Paraphrasing Albert Einstein, not to stop questioning is the human characteristic on which all progress is based. Note that in this book, "intelligence" only appears sometimes as the adjective "intelligent," and just for signaling decisions leading to actions beneficial for both the person and society. Intelligence is a concept referring to apply rationality in the best form possible and, for instance, if the capability for computing were considered an important attribute of intelligence, today it is by and large surpassed by computing machines. The great technological problem is how to endow computers with rationality, with *Logos*.

7.3. Referring to human rationality requires jointly considering conceptualization, or categorization, and analogy, because both permeate language and reasoning. If concepts are abstract products of thought, analogy is what allows their linking, the establishment of relations between them that is the essence of reasoning. But neither is reasoning just analogy, even if analogy seems to play in it the character of an unavoidable natural mechanism, nor can reasoning be identified with "thinking" that comprises aspects only indirectly related to reasoning, for instance, imagining and remembering, to say nothing of the sometimes relevant influence emotions and previous beliefs have on reasoning. In any case, memory or storage is basic for both conceptualization and analogy, that is, for organizing in the form of concepts or categories what is being known, and for establishing some relationships with what was known earlier and with the external reality.

Without some relationship to reality, knowledge risks being insufficient for both acting and taking decisions; additionally, it is important to be conscious that any current knowledge, at least of the world, is not forever, but that it has a caducity date that is not previously known as it is with medicines. Knowledge is not preserved like medicines in a closed container; knowledge is not fixed and isolated; it is always in flux; it varies with time.

The often-repeated linguistic expressions "of the same kind of", or "like that", already announce in language the relevance of analogy in creating "categories" or "collectives", naming concepts, the basis of, at least, not spontaneous reasoning. It seems that comprehending something new lies in the possibility of comparing it with what is already comprehended, or believed to be so; namely, in making analogies with past experience or with already acquired knowledge. Denoting or labeling concepts, or categories, by words, that is, by "naming" them, is essential for remembering, managing, and relating concepts; naming nonempty categories is an efficient way to help people's reasoning.

For comprehending a new concept, or to find a new one, or to pose a question, or to capture some unobvious relationship, the analogy with something previously known or firmly believed seems to be a step forwards either good enough, or maybe necessary; it is, perhaps, for this reason that analogy plays so an important role in questioning. Posing questions, and guessing their answers through analogous former situations, are typical expressions of rationality. At the end, in a rough summary, science consists in a specialized and ordered guessing in particular domains, aiming at solving previously identified problems whose solution could be stated in the conditional form "*provided* this, *then* that". Foreseeing, essential for surviving, is almost always done in conditional terms; for instance, weather forecasting is always done under the antecedent of what is currently known about the atmospheric situation, and it is due to the antecedent's variation that weather forecast failures lie.

Thinking is a wide natural phenomenon whose functioning, produced by neural connections in the brain, still remains partially unknown, and neuroscientists are today searching for the role language plays in reasoning. Actually, and from what has recently been discovered, it seems that relational thinking is what supports natural reasoning jointly with how emotions, intentions, and desires can direct it, and even if forms of spontaneous thought are being explored. For many years, thinking has been recognized thanks to some of its signs such as, in a first instance, the capability of organizing knowledge in categories, of comprehending and managing conditional statements well, and the linking of categories to reach conclusions, reasoning, in sum.

Hence analogy is an important subject to reflect on how concepts are generated, on how they can be scientifically domesticated, and on how analogy initially helps thinking with the kind of organization in which reasoning consists. In this respect the following imaginary example can be in order.

7.4. In the beginning, when primitive people started to manage stones for building walls preserving them from predators, they, at least in some way after several trials, empirically realized that stones were of different forms, sizes, and weights (even without having these concepts/words), and that it was better to place those that were bigger in size and height in the wall's first level for grounding the wall properly in the land. In this experiential way, and after trial and error, the concept of "big" was born, and it soon was applied to materials other than stones, such as pieces of wood obtained from trees, as well as to the same trees, walls, animals, and the like. The category *big*, referring to stones, walls, trees, animals, and the like, could then be managed to facilitate communication. That is, the meaning of "big" appeared for the first time allowing the people to say, for instance, that some lake was big by comparing it with others that were able to be skirted around in fewer days; possibly, this also led to taking into account the concept of the "size" of lakes, trees, animals, hills, and so on, and the new and bigger category *size* was born. Possibly, also the qualitative concept of the "weight" of stones, animals, pieces of wood, and so on, jointly appeared; it only turned to be quantitative when, fueled by trade, the first mold for grains, or the first balance, appeared.

It seems that concepts don't appear isolated from each other, but linked in the form of chains of concepts keeping between them some empirically perceived relationships. Maybe some concepts, appearing from the direct management of physical objects, are either followed, or intermingled, with other concepts coming from the management of the former ones; this can be, for instance, the cases with *length* and *weight* if derived from the pair *big* and *small*, and allow the introduction of "large" and "middle" size, for instance. In ways like those, the use of the concept *big* spread all over language thanks to the analogies established by comparing between them several different families of objects, and, perhaps, also originating other concepts such as *size*. Furthermore, and perhaps by comparing with the naked eye the height of trees, hills, and mountains, the concepts of *high* and *top* appeared, in short, playing some Wittgensteinian sort of language game with words.

It seems difficult to believe that a concept could appear isolated from others, and it is more credible that concepts are generated inside a more or less organized conglomerate of old and new concepts that migrate between several universes of discourse. If it is very difficult, if not impossible, to establish the genealogy of a given particular concept, what seems without doubt is that its genesis is grounded on experience and analogy; once experience is presupposed, rationality is not even imaginable without the capability of making analogies, of appreciating some family resemblances. It is like in the movie *2001: A Space Odyssey*, in the beginning of which the monkeys imagine that some bones can be used for fighting against their enemies by beating their heads the way they beat a stone, and showing how the first weapon appeared.

Very often, something is considered to be understood when both the similarities and dissimilarities it shows with something else previously known are captured; the roots of understanding and creativity are immersed in analogy. The perhaps false story of Newton's apple, and also that of the Einstein lift, are but examples of it.

Those simple and imagined cases with primitive people just try to illustrate how categorizing could empirically have been started and, once connected by analogy, fuel reasoning. That is, helping the acquisition of reasons/arguments enough for acting or deciding to do or not to do something, such as to place a settlement at the border of a big lake and at the top of a hill, for counting with both water and fish, and the possibility of observing, with enough time to react, if dangerous animals, potential enemies, or just unknown people, approach the settlement, as well as for doing the necessary constructive actions for building such a settlement, in sum, constituted conjecturing or refuting the adequacy of placing the settlement there. Note that, in the first place, it implies managing conditional statements well such as,

"If the settlement is placed here, then it is possible to foresee the danger that could come from an unexpected approach," not implying, of course, the impossibility of placing the settlement elsewhere. Indeed, conditional statements play a central role in reasoning.

For mentally establishing a commonly shared concept, a previous group's experience with elements of some first family of objects in a universe of discourse interesting for the group is required; later on, the newly acquired concept migrates to be applied in other universes by analogy. Hence, it is important trying, more or less formally, to consider as an initial point what concerns the meaning's representation of imprecise concepts (such as they are "big", "high", etc.); they actually permeate plain language and ordinary reasoning. The goal of its scientific domestication as done before, is actually relevant. Note that concepts are singularized, even isolated among them, and also remembered, thanks to their names; they are commonly recognized, jointly with their opposites and negations, thanks to the names, or linguistic labels, or predicates, that once stated, help to avoid the confusion their mixing could produce. Predicates P, through the elemental statements "x is P", allow capturing the meaning of concepts; it is in this sense that each predicate can be seen as the "mother" of a corresponding concept, and that overlapping of meaning and synonymy are seen as important linguistic phenomena.

7.5. Either a kind of Platonic precedence of concepts over previous experience with physical or virtual samples for their application, or the possibility of a strictly private reasoning on them, before their naming or their inclusion in language, is something illusory and always dangerous. Concepts can only be communicated after counting with an initial meaning. The before-mentioned elemental statements "x is P", once intercommunicated among a group of people, are what allow the common comprehension of the concept, and its use in complex statements, after the group's members share some common meaning of them. Only after some concepts can be managed by recognizing "this is that" can abstract reasoning start. There cannot be human rationality without meaning; semantics and rationality are strongly intermingled. Semantics, the essence of understanding, comes from observing and analyzing what is external to people; from, for instance, distinguishing between a lake and a river by describing, after being acquainted with them, what is a lake and what is a river, which characteristics distinguish one from the other, passing, for instance, from "these lakes" to "lake".

Human beings need intercommunication, sharing concepts and reasoning, accepting or denying them; they are social beings who are not only directed by genetics but, also, by a common culture of which language is a manifestation helping communication by designating concepts by words that allow us to share, modify, and improve them. Each concept charges with a cultural back for the members of a more or less culturally homogeneous group of people, and it is often modified through the typically social experience provided by talking; for instance, as neuroscientists observed, it is easier to reason with familiar than with unfamiliar concepts. In the case of an unfamiliar concept, analogy permits a first approach to what it means.

It should be pointed out that almost all those concepts, if not all of them, are of an *imprecise* character; that is, they cannot be expressed/captured by *if and only if* definitions, as they are the *precise* concepts of mathematics, for example, "prime number", "continuous function", and so on.

Nevertheless, concepts such as the mathematical ones only can appear once the capability of *abstracting* is acquired from experience, and from something stored in the memory with the aim, for instance, of clarifying a subject through its representation in a formal setting, and perhaps for being able to compute some of its aspects with the help of an artificial language based on some new but related and stringently defined precise concepts. This is possibly how the surveyors at the Nile in ancient Egypt went from the cord with 3, 4, and 5 knots to mark rectangular triangles in the land to Pythagoras' famous theorem. Note that theorems are, essentially, conditional statements the validity of whose consequent depends on that of its antecedent; for instance, Pythagoras' theorem holds on a Euclidean plane but not on a spherical surface. Indeed, in some way it "defines" the Euclidean plane.

In any case, the way leading from ordinary to abstract-formal reasoning is a long one, even if it were noticed that the abstraction capability is manifested from the very beginning, because without it neither imagining nor understanding a conditional statement nor that of making any analogy is actually possible. Abstraction and imagining seem to be the brain's capabilities developed, for instance, through mental processes such as that of passing from "this stone" to (the category) *stone*, and that in no way is limited to formal abstraction. For instance, just arriving at the category *bird*, and stating, "Almost all birds fly," already requires some abstraction; without a minimum capability of abstraction, reasoning does not seem actually possible.

7.6. All that jointly proceeds with the human capability for questioning; posing and trying to answer questions seems to be the reasoning task. In this sense, reasoning can even be seen as a "mechanism" directed to answer questions. It is for this that *Homo Sapiens* could be better known as *Homo Quaerens* (the seeker and the sought), the current *Homo* species able to pose questions expressed by words. There are other animals that also "know", but they seem unable to pose questions to themselves, and especially complex questions articulating chains of simpler questions; questions coming from a deep curiosity seem to be a strong intellectual engine of reasoning. Without categorization, analogy, abstraction, and good management of conditional statements, complex questioning coming from dynamical situations does not seem actually possible, at least in forms intellectually fruitful, whose answers can allow foreseeing and surviving in a partially unknown environment, or concerning difficult problems for whose solutions good guessing, and some suitable kind of proving conclusions, is necessary.

Language, abstraction, and questioning are what facilitated our ancestors in arriving at fiction and especially at "telling fictions", and at "talking on fictions", or at "writing fiction", through imagining virtual, not always properly physical, entities and situations. These imaginations, sometimes transformed in myths, are what allowed Homo Quaerens, over the last 70,000 years, to cooperate effectively by associating a number of them surpassing the small number of those that can be directly acquainted with some concepts, questions, guessing, and answers. That does not seem to be much greater than around 100 people, a more or less big herd. In such a way, commerce, religion, law, social and political organizations, science, and especially the capabilities of transmitting information and foreseeing, were born. This seems to be the cultural evolution giving to Homo Quaerens its singular and marvelous capability for questioning and answering with fruitful conjectures and, in consequence, innovating and even showing creativity. Without such capability, the species surely would either disappear, or remain bound into regions where predators were kept under sufficient control, with food and water enough to survive, more or less what currently happens with other mammals, gorillas and chimpanzees that, for instance, neither spread around the planet, nor show but a minimum innovative capability, nor are able to write; they seem to lack the word, although they can learn some words once a person teaches them.

Along the intellectual processes leading to know deeply and to innovate, it cannot be forgotten that recognizing contradictions also plays an important role; if it can be said that a person grows intellectually through successive infections of contradiction, it is clear which benefits recognizing contradiction facilitate mankind, and which advantages of escaping from contradiction add to Homo Quaerens' life; there is no doubt that the appearance of a contradiction makes mentally sane Homo Quaerens stop any reasoning and that it seems to be a central part of an abstract formal study of reasoning. When the Pythagorean philosophers believed that the length of the diagonal of the square contradicted their belief that it should be a fraction of its side's length and that, consequently, such length cannot be a number, they just hid the result, and it took many centuries to recognize that "irrational" entities such as $\sqrt{2}$ are but numbers.

The history of mathematics, and also that of science, contains many cases in which the final task was to surpass (apparent) contradictions by constructively creating new formal entities, new symbols endowed with a new meaning, permitting the guessing of new questions, and opening new frontiers to research. Because without real and complex numbers, differential and integral calculus would not exist, current science is indebted to the passing over of rational numbers done by introducing the irrational ones. It can be said that if the positive integers are "found" in the world and were abstracted by the necessity of counting, the other numbers are a human creation that needed former knowledge and abstraction, some previous mathematical operations, and representation frameworks for it, for instance, quotients for the rational and square roots for the first irrational numbers, to say nothing of transcendent numbers such as π and e. Without formal abstraction numbers would not be known. It is nevertheless obvious that people don't like to continue when actual, or believed, contradictions appear; everyone considers that contradiction is not acceptable.

What marks the basic jump in the way to arrive at formal abstraction is the *representation,* in a suitable framework, of concepts by symbols endowed with meaning. Representation is a concept older and wider than in a mathematical setting alone, as shown, for instance, by the old representations of animals and people in the pictures on the walls of caves or, in a conversation, drawing in sand a shape by helping to graphically show some aspects of something. But a representation's frame, be it a wall or the sand, is always necessary.

Representation is inherent to reasoning, and the formal case itself requires several previous analogies, abstractions, managing of conditionals, and formal constructions; for instance, when representing a sinusoidal curve in a Euclidean plane, even if it is a smudged one, the mathematical plane should have been previously constructed by abstraction of an analogy with a flat surface, and the name of the function by an abstraction by analogy with a physical figure, and so on. It does not seem that a so-relevant concept for "reasoning on reasoning" as is that of *representation,* could be deeply analyzed without, at least, taking into account *meaning, conditionality,* and *analogy.* The possibility of "representing" is crucial for articulating ordered reasoning, and representing something by symbols abstracting some features of the real things is still more important.

Additionally, it should be pointed out that even based on analogy, reasoning is not a uniform mechanism but one that specializes itself by attending each wished goal. For instance, the deductive reasoning of the mathematical proof is a specialization, done in a previously constructed formal framework and with an invented artificial language, directed to check that either a conjecture, or a refutation, can be actually accepted in the corpus of mathematics, and defining the inferential relation through a special interpretation of the meaning of conditional statements. Reasoning continuously refines itself by more reasoning, leading to the adoption in each case and through several abstractions of some special organization that allows the advance of knowledge on the territory to which it is applied, such as with reasoning on reasoning.

This view is a good reason for, directly or indirectly, trying to learn from a first-class mentor when, for instance, either one decides to prepare a PhD dissertation, or to become a sculptor, a painter, a writer, a musician, a singer, and so on. In all of these cases, and apart from learning the necessary technicalities, there is the strictly important aspect of learning how to recognize "good" questions, the form of actually facing them, looking for antecedents further than technicalities but in the context in which questions are posed, seeing interesting analogies, figuring out their answers, capturing when an answer could be seen as possibly fertile, and the like, in sum, by acquiring what is to count with a clinical eye on the corresponding topic.

All that requires big doses of analogy, which is only possible to learn in close contact with the best real praxis possible. A "good office" is always acquired by intensively working as an apprentice in the workshop's team of a great master artisan, seeing many times the difficulties appearing in both the design and along the process for arriving at the end, and, especially, keeping permanently excited, worried, and critical for all them. After graduation, specialized forms of reasoning require an apprenticeship period; good researchers fight with, and against, the work of the greatest scientists. Parodying the words of the great geometer Luis A. Santaló, a scientist only can be considered so if, from time to time, he or she publishes original, new, and interesting results; at least those that are neither original, nor new, are useless; they don't contribute to advance knowledge or practice.

7.7. With all that has been said, there lacks the addition of a short comment on formal deduction, that is, on the deduction done in a formal framework, the reasoning whose conclusions, or consequences, are attainable by means of algorithms reproducing, step by step and without any jump, the passing from a statement to the next one by using the rule of modus ponens.

As said above, this type of reasoning is the safest one, but it does not mean that everything that can be posed in a formal framework can be attained by an algorithm, as shown by those mathematical problems having been formally proven to be undecidable, that is, impossible to neither prove them nor their negation. In those problems and in each case, some solution can be envisaged by guessing at it, but it cannot be reached deductively. Not all that is thinkable has, notwithstanding, the possibility of being deductively found and hence for the advancement of knowledge, guessing is necessary; without conjecturing and refuting, without guessing, no knowledge seems to be possible. The human mind's capability of reaching solutions by deduction is, sometimes, blocked; something that forces us to be humble, and paraphrasing Albert Einstein, accepting that what is actually marvelous is not only formally to deduce some conclusions, but also formally to prove deductively that there are cases in which this is not possible.

For all that, some kind of black over white symbolic representation can be necessary, but it should be taken into account that, as with visual monochromatism, intellectual monochromatism is dangerous for interpreting realities inasmuch as things are seen in color and, at least, greys are important. Not only is gradation basic, but considering the particular situation in which something is going on is essential when attempting to construct a model for it.

The kind of intellectual achromatopsis consisting in reducing situations to only a model in black over white, can lead to simplifications that do not allow us to appreciate deeply all that is actually there, and can limit the model's usefulness; this is what Lotfi A. Zadeh tried to avoid in 1965 with the introduction of fuzzy sets.

Nevertheless, starting from a simplistic black over white description can sometimes help to pose some problems with an initial and sufficient clarity for a subsequent advance of thought.

References

1. D. Hoffstater, E. Sander, *Surfaces and Essences: Analogy as the Fuel and Fire of Thinking* (Basic Books, New York, 2013)
2. M. Black, *Models and Metaphors* (Cornell University Press, Ithaca, 1962)
3. E. Trillas, J.M. Terricabras, A scrutiny in representing meaning and reasoning, *Archives for the philosophy and history of soft computing*, vol. 1 (2016), pp. 1–70
4. L. Wittgenstein, *Philosophical Investigations* (Basil Blackwell, Oxford, 1958)
5. B. Russell, *The Problems of Philosophy* (Williams & Norgate, London, 1912)
6. G. Shafer, *The Art of Causal Conjecture* (The MIT Press, Cambridge, 1996)
7. S. Watanabe, *Knowing & Guessing* (Wiley, New York, 1969)

8. E. Trillas, I. García-Honrado, Unended reflections on family resemblances and predicates migration, in Proceedings of EUSFLAT (2011), pp. 598–606
9. E. Trillas, How does science domesticate concepts? in *Archives for the Philosophy and History of Soft Computing*, vol. 2 (2014), p. 12
10. L.A. Santaló, *La educación matemática*, hoy edn. (Teide, Barcelona, 1975)

Chapter 8
A (Naïve) Symbolic Model of Ordinary Reasoning

For dealing with what can be of interest to someone, and usually directed to action, or to reaching some goal, or to understanding, or to foreseeing, and so on, the natural instrument of thinking facilitates the linkage of concepts. Concepts originate either for clarifying something, justifying some previous action, or foreseeing how to reach some wished goal, and the like. Hence, at the bottom of reasoning, and in whatever form it can be shown, abstraction, categorization, analogy, and meaning play pivotal roles.

Ordinary reasoning, lying in the natural phenomenon of thinking and consisting in several modalities, is in itself difficult to be formally defined and analyzed, however, by researching what it does instead of what it is, the parts ordinary reasoning consists in can be distinguished, and even some hints on creative reasoning become possible. At the least, this operational strategy allows the construction of an initial symbolic model of reasoning that, perhaps, could be useful for helping us look at how to reach higher degrees of "machine rationality" by applying it to some particular contexts. What will remain debatable and actually open, is whether the logos a computer could be endowed with can become like a "total" human Logos.

It should be noticed that, essentially and as has been said before, reasoning is impossible without counting on some previous information of its particular subject.

Without perception, recognition, distinction, and abstraction, neither conceptualization, nor analogy, nor the linking of concepts, seems to be possible. A person is a very complex physical entity endowed with many sensors, and the interpreting, representing, controlling, and reasoning "machine," the brain, is actually also of enormous complexity and facilitates her or his view of the world. All this is but a small part of the important challenge of knowing how the brain's complex and organic machine actually works, and currently almost the only ones of which that can attempt to submit to some level of mathematical modeling are plain language and formal reasoning. It will probably be many years before autonomous computers can come to "see" the world and, consequently, represent it.

E. Trillas, *On the Logos: A Naïve View on Ordinary Reasoning and Fuzzy Logic*, Studies in Fuzziness and Soft Computing 354, DOI 10.1007/978-3-319-56053-3_8

Reasoning allows either rejecting that which contradicts some initial information on something, or considering what is not in contradiction with it. It is not possible to reason on nothing; from nothing only nothing can be reached; a brain (either human or hardware), empty of recognized concepts and relationships, can do nothing concrete, and reasoning basically consists in conjecturing and refuting from some previous information, evidence, or knowledge. Any reasoning departs from something that is physically seen or mentally believed, known, or just accepted; some categories, collectives, concepts, and relationships among them are at the beginning of all reasoning whose prosecution is done through a linkage of the items of such starting information, and limited by what can be presumed on it. Meaning, as well as the chaining of shared meanings, is at the roots of reasoning.

8.1. Consequently, towards a formal analysis of reasoning, it should first be stated what "that follows from this" means, and "this is in contradiction with that", and by also including when something can be contradictory with itself, self-contradiction. For instance, in ortholattices, self-contradiction is just limited to its minimum 0, but not in De Morgan algebras and algebras of fuzzy sets; self-contradictions also appear in plain language.

After the basic linking of concepts, the idea of contradiction is crucial for concluding something that, more often than not, will be of a "provisory" and unsafe character. In addition, it should be pointed out that, as has been said, the conclusions reached in ordinary reasoning have unknown data of caducity, not always due to the disappearance of some initial information, but to the appearance of a new one. Information is not fixed; usually, it flows continuously.

Let's suppose the existence of a (primitive) symbolic relation $\leq$ between words, collectives, or concepts, being but a naïve mathematical representation of the "intellectual" relationship of "inference" between them. With it, denote

- $p \leq q$ shortens "q is forward-linked with p", or, written $q \geq p$, "q is backwards linked with p", and regardless of how the link is actually established
- p^a shortens "antonym of p", and p' shortens not-p
- $p \not\leq q$ shortens "q is not forward-linked with p", and does not always imply $q \not\leq p$

with p, q, and so on, representing words, collectives, concepts, composed concepts, or relationships among them and that, with the aim of simplifying and regardless of which role language effectively plays in reasoning, they are denoted by the corresponding words naming them "predicates", "linguistic labels", or "statements" for short.

The relation $\leq$ is taken as a "primitive" one between words; the graph constituted by the statements into consideration and the relation $\leq$, is the basic ground on which reasoning is represented. With $p \leq q$ it is simply stated that an intellectual "movement" from p to q is recognized, and with $p \not\leq q$ that such a movement is not recognized.

In principle, there are not supposed more specific properties for the relation $\leq$ further than it is not empty, that there are actually pairs (p, q) linked by it, or verifying $p \leq q$, that pairs of concepts effectively linked exist. A possible name for $\leq$ is "*relation of natural inference*" allowing to change $p \leq q$ as *q is naturally inferred from* p, and provided it were $p \leq q$ and $q \leq p$ simultaneously, it would be said that p and q are *inferentially equivalent*, and shortened by $p \approx q$.

Note that there can exist pairs of statements p and q such that it is neither $p \leq q$, nor $q \leq p$ such as "John is now in the swimming pool," and "John is now flying in a plane"; these pairs are inferentially not-comparable, or inferentially isolated pairs, and it is written p NC q, shortening "inferentially not-comparable" by NC. The relation $\leq$ is not always linear.

Hence, given two statements p and q, it just can be either $p \leq q$, or $p \geq q$, or both and then $p \approx q$, or p NC q. It is always supposed that it is $p \leq p$, that each p is linked with itself; it cannot seem a rare presumption, and excludes the possibility that $\leq$ can be empty.

The relation $\leq$, here taken as a naïve and primitive concept, tries to represent the linkage, or intellectual movement, people establish between words, concepts, or collectives by usually connecting their meanings, and with which reasoning seems to be possible. Additionally, it is supposed that such a relation $\leq$ is a common one for all mentally sane people. For instance, in formal classical logic on Boolean algebras, $p \leq q$ comes from the understanding that it asserts the conditional statement "if p, then q" that, once it is made equivalent to the affirmative statement "not p, or q", and supposing it is "a tautology" (equal to the greatest element of the corresponding Boolean algebra), gives the natural (partial) order $p \leq q \Leftrightarrow p \cdot q = p \Leftrightarrow p + q = q \Leftrightarrow p' + q = 1$, of the Boolean algebra. To formalize reasoning mathematically, the previously "undefined" relation $\leq$ should be suitably selected, as is done in the former Boolean case.

Nevertheless and right now, very few properties of the inference relation $\leq$ are presumed with the goal of keeping a very general framework for describing reasoning under a minimal number of constraints but, when necessary for either adapting it to a particular setting or for obtaining additional results, some additional properties are added. This is done in the line of not considering more than what is strictly necessary for posing some relevant questions (the old Occam's razor methodological rule of the fourteenth century), but not less than what can allow reaching something of interest (the twentieth century's Menger addenda to the razor).

By $p \leq q$ is simply indicated that q is inferred from p, but neither how it is done, how q is effectively reached, nor if there are intermediate steps for it; later on, it is made explicit for some particular, specialized, and formalized kinds of reasoning. In some sense, we take care of nothing less than the existence of a general "intellectual link" between p and q expressed by $p \leq q$, whatever it can be, or how it can be understood and represented. No methodology for reaching q from p is attended right now, and it is just supposed that q is inferentially linked with p in a naïve form; that a movement from p to q exists.

It is just presumed that jointly with $p \leq p$, it always holds, as has been formerly said,

$$p^{a} \leq p',$$

coming from many examples, but whose reverse does not always hold, and showing that the negation is an inferential upper limit, often inaccessible, of the opposites; that is, that p^{a} is not inferentially equivalent to p' except in those irregular linguistic cases in which there is no opposite p^{a} of p in language, and p' is taken instead of p^{a}.

Because no one will recognize validity and usefulness to self-contradictory statements such as "The white carpet is black," the worst foil a statement p can show is to be self-contradictory, that is, to verify $p \leq p^{a}$, or $p \leq p'$. Note that, provided the triplet (p, p^{a}, p') were $\leq$-transitive, that is, from $p \leq p^{a}$, and $p^{a} \leq p'$, it would follow that $p \leq p'$; the first relation implies the second but not reciprocally. Anyway, as shown, the triplet's transitivity cannot be generally supposed, and its necessity or sufficiency should be checked and eventually stated in each case. Analogously, the worst foil a pair of statements p and q can show is to be contradictory to each other; that is, $p \leq q'$, or $q \leq p'$.

8.2. The (available) previous information is usually presented by a set of linguistic statements $P = \{p_1, p_2, \ldots, p_n\}$, called the set of premises, that can be compacted in a single statement (p) by a sequential conjunction of the premises once they are numbered according to some previously accepted criteria related to the subject under consideration. Note that, actually, it does not seem possible to fix a universal rule for always ordering the premises, but that it is privative of the context in which each set P is inscribed. This conjunction p tries to sum up the information conveyed by P, and is called the *résumé* of the premises; for instance, with $n = 5$, it will be

$$p = (((p_1 \cdot p_2) \cdot p_3) \cdot p_4) \cdot p_5,$$

obtained through the ordered sequence of the four successive conjunctions

$$p_1 \cdot p_2 := p_{12}; \quad p_{12} \cdot p_3 := p_{123}; \quad p_{123} \cdot p_4 := p_{1234}; \quad p_{1234} \cdot p_5 = p,$$

each giving a (new) statement.

Of course, provided it could be supposed that the conjunction $(\cdot)$ is associative and commutative, p could be simply written

$$p = p_{k1} \cdot p_{k2} \cdot p_{k3} \cdot p_{k4} \cdot p_{k5}, \text{ for any permutation } (k_1, \ldots, k_5) \text{ of } (1, \ldots, 5).$$

Nevertheless, in language the conjunction's commutative and associative properties cannot always be presumed, because usually time intervenes and makes it unreal; for instance, "She entered the room and start crying" and "She start crying and entered the room" or, in a joking fashion, "He was judged and hanged" and "He was hanged and judged." With respect to the associative property, it should be just observed that it cannot always be taken for granted because, for instance, people often distinguish $p \cdot (q \cdot r)$ from $(p \cdot q) \cdot r$ as soon as time can be seen coincidental for q and r, but not for p and q; the associative law is a syntactically originated rule in mathematics and that facilitates the calculus, but in ordinary language commas are very important for a good understanding of texts and either "avoiding parenthesis or commas", or "moving commas", is not always acceptable in all writing. It can lead to changing the meaning of the corresponding statement.

Hence translating into reasoning and language the typical properties of mathematical structures, such as those holding in a lattice, cannot always be done. For instance, the usual supposition that the lattice's conjunction translates the linguistic conjunction is very risky by implying that the linguistic "and" is commutative and associative, and coincides with an, perhaps not actually existing, upper limit of all the statements from which the two members of the conjunction can be forward inferred. In this respect, it should be recalled that in the basic algebras of fuzzy sets, the greatest conjunction is that given by the lattice operation min between fuzzy sets, the only case in which, were the disjunction additionally taken as the lattice operation max, the algebra is a lattice; it is but a limit case.

It should be pointed out that provided the operation of the conjunction $(\cdot)$ were to verify the usually accepted two properties expressed by

$$p \cdot q \leq q, \quad \text{and} \quad p \cdot q \leq p,$$

the résumé of P would verify $p \leq p_i$, for all i between 1 and n. For instance, with $n = 3$, it is both $p = (p_1 \cdot p_2) \cdot p_3 \leq p_3$, and $p \leq p_1 \cdot p_2$, and hence $p \leq p_1$ and $p \leq p_3$, provided the triplets $(p, p_1 \cdot p_2, p_1)$ and $(p, p_1 \cdot p_2)$ were transitive. That is, under transitivity suppositions, all premises are inferred from its résumé; in such conditions, the résumé compacts and "explains" the premises (they are obtained by a forward link from the résumé). It should be pointed out that to state that each p_k follows from p, transitivity has been accepted to hold, is something that cannot always be presumed.

In what follows, presuming P is without contradictions, and because if p is self-contradictory P cannot be safely taken, it is cautiously supposed that neither the résumé p of P, nor its negation p', are self-contradictory; that it is $p \not\leq p'$, and $p' \not\leq p$, for which it obviously suffices to have $p \not\leq p^a$ and $p' \not\leq p$. Neither is there a way for forward inferring p^a from p, nor p from p'. It should be pointed out that in the limit case of Boolean algebras, $p \leq p'$ and $p' \leq p$ are, respectively, equivalent to $p = 0$ and $p = 1$.

8.3. Once this reasonable condition on p is accepted, it is said that q refutes P, or that q is a *refutation* of P, if and only if

$$q \leq p^a, \quad \text{or} \quad q \leq p',$$

that is, either the opposite or the negation of p can be inferred from q. Note that inasmuch as $p^a \leq p'$ is always supposed, provided the triplet (q, p^a, p') were transitive, $q \leq p'$ would follow from $q \leq p^a$.

Those q that are not a refutation are a *conjecture* from P: $q \not\leq p'$, where it sometimes suffices to have $q \not\leq p^a$. For instance, on the conditions of a currently extremely dense cloudy atmospheric situation, it can be conjectured that it will rain soon, but refuted that it will be sunny in one minute.

Note that, once P and p are given, there are no involved statements other than refutations and conjectures; and also that once $p \not\leq q'$ is known, it can only be either $p \leq q$, or $q \leq p$, or p NC q. Hence conjectures can be classified in those verifying, respectively, $p \leq q$, $q \leq p$, and p NC q, respectively called *(ordinary) consequences* of P, *hypotheses* for P, and *speculations* from P.

Note also that ordinary consequences (nothing else indeed than the conclusions by the primitive notion of forward inference) can be considered both the basic treat of reasoning, and the form for developing what is hidden in the résumé; without natural deduction, the search for ordinary consequences by forward linkage, reasoning could not exist. Natural deduction appears as the basic brick of reasoning; the formal methods of deduction, considered later, are but specialized forms of deduction done in a formal framework.

It should be pointed out that "this" natural deduction should not be confused with the methods of formal deduction in classical logic, and as it is in, for instance, that of Gentzen, implicitly supposing all the laws of a Boolean algebra are also called <natural deduction> but are actually for deploying, step by step, formal deduction. Ordinary deduction is weaker than formal deduction; the formal forms of deducing are but restricted specializations of the ordinary one submitted to ruled processes for arriving at a conclusion.

Under a few minimal hypotheses on the negation and the inference relation $\leq$, those q such that $p \leq q$, the *ordinary* consequences of P, can be proven to be conjectures; that is, such that $p \not\leq q'$. Such hypotheses on the negation are the following two,

- $p \leq q \Rightarrow q' \leq p'$ (the negation reverses linkages)
- $q \leq (q')'$, for all q (the negation is "weak")

and for the inference relation $\leq$ is supposed to hold the transitive property for the triplets (p, q, r) under consideration; that is,

$$p \leq q \quad \& \quad q \leq r \Rightarrow p \leq r.$$

With these three properties, of which only the first is always acceptable as a fix property of negation,

- If it were $p \leq q'$, it would be $(q')' \leq p'$ and, from $q \leq (q')'$, $q \leq p'$ would result and, hence, $p \leq p'$ from the hypotheses that q is a consequence, $p \leq q$, and that the triplet (p, q, p') is transitive. An absurdity is reached, because it is supposed $p \not\leq p'$; thus, it is $p \not\leq q'$.

Thus, under the above three suppositions, the consequences q inferred from p are conjectures and q can be characterized by $p \leq q$ without adding $p \not\leq q'$. In this form (ordinary) consequences appear as particular cases of conjectures, as some British thinkers of the twentieth century believed it should be, and ordinary deduction appears as a modality of conjecturing. Nevertheless, it should be pointed out afresh that neither the weak nor the strong character of negation, nor transitivity, are always presumable; for instance, p and $(p')'$ are often isolated. The analysis of ordinary reasoning without transitivity, and, further that reversing $\leq$, an additional property for negation, is an open subject.

Additionally, no two consequences of P can be contradictory, because if it were consequences q and r such that $q \leq r'$, from $p \leq q$, and supposing the transitivity of the triplet (p, q, r'), it would follow $p \leq r'$, that, being $p \leq r$, and hence $r' \leq p'$, the supposed transitive triplet (p, r', p'), implies $p \leq p'$, which is absurd; of course, and by the same argument, no consequence can be self-contradictory. This is in agreement with what is, at least, expected on deduction, and notice that were the negation strong $(q \approx (q')')$, as sometimes it is in language, a fortiori this result would also hold. Nevertheless, because in ordinary reasoning it is often the case of reaching contradictory consequences, it means that either such a particular reasoning is not deductive, or that transitivity fails in it.

On the above hypotheses, conjecturing is more general than natural inference; the processes for obtaining consequences are those of *deduction.* Nevertheless, it should be pointed out that these conclusions hold under the above hypotheses, and that their lack can cause their failing; in any case, what has been established cannot be taken for granted by any reasoning. Even a soft formalization such as the one here is but a model only applicable under what is presumed for building it. Note, in this respect, that of the above three hypotheses, only the first (negation reverses $\leq$) always holds, but that negation is weak or strong depending on the particular form of negating, and that some transitivity does hold cannot always be taken for granted. Anyway, because through ordinary inference contradictory conclusions are often reached, it at least reveals that neither the weakness of the form of negation nor the transitivity of the linkage can always be presumed to hold. Correct forward inferencing, ordinary deduction, seems to need the above hypotheses.

Another type of conjecture is those q $(p \leq\!\!/\, q')$ that, being not self-contradictory $(q \leq\!\!/\, q')$, are such that it is $q \leq p$, and they are called *hypotheses,* or *explanations*, for P; they are reached backwards from the résumé of P. Because p is forward inferred from q, p is an ordinary consequence of q, and provided it held the property $p \leq p_i$ $(I = 1, 2, \ldots, n)$, with which from $q \leq p$, and also transitivity, it would follow $q \leq p_i$, all the premises would also be inferred from q. This is the sense under which a hypothesis explains all the premises; something that can fail if transitivity fails.

What cannot be proven is the inexistence of pairs of different and contradictory hypotheses, something in agreement with the praxis; it is not rare to have two different and contradictory hypotheses for explaining something. Note that if the résumé p is, in turn, a consequence of the hypothesis q, the only hypotheses that can be consequences of p are those inferentially equivalent to p, inasmuch as they should verify $p \leq q$ and $q \leq p$. Hence the statements inferentially equivalent to p can be avoided because, inferentially, they are nothing else than p, and taking p as a hypothesis for P is to say that p explains itself; p is never taken as a hypothesis for P. The processes for obtaining hypotheses are those of *abduction*. In principle, $q \leq p$ without $p \leq q$ seems to show q is "simpler" than p, that the explanation is done by avoiding something that is included in the résumé p.

The remaining class of conjectures, those q that are neither a consequence, nor a hypothesis, should be conjectures isolated from p, that is, verifying p NC q. They are characterized by verifying p NC q, jointly with either $q' \leq p$ or p NC q', because q cannot be a refutation ($p \leq q'$), and $p \leq\!/\, q'$ is equivalent to $q' \leq p$ or to q' NC p. Hence there are two types of these conjectures,

- Those characterized by p NC q and $q' \leq p$
- Those characterized by p NC q and p NC q'

known as *speculations* from P and called, respectively, of type-one and type-two.

Because $q' \leq p$ implies $p' \leq (q')'$, the double negation of a type-one speculation can be forward inferred from the résumé's negation even if q is isolated from p, but provided the linkage $q \leq (q')'$ were to hold, or that also $(q')' \leq q$ hold (that q and $(q')'$ are inferentially equivalent, and the negation is "strong"), then q would be inferable from p'. Under these types of negation, all type-one speculations are inferable from the résumé's negation; to some extent, type-one speculations are deductively reachable. For instance, if the negation is weak, then q could be backward reached from $(q')'$, and provided it were intuitionistic, $(q')' \leq q$, q could be forward reached from $(q')'$. A type-one speculation q is not deductively reachable from the résumé p, when it is q NC $(q')'$, when q and $(q')'$ are inferentially isolated.

Nevertheless, all type-two speculations are inferentially wild conjectures, because they and their negations are inferentially isolated from the résumé p; type-two speculations cannot be inferentially reached from the résumé or from its negation, and obtaining them corresponds with a *creative* reasoning, opening the idea of what is not directly inferable from p, or from p', and could mean something new. It can be said that the search for this type of speculation is, properly, an *inductive* reasoning, or an *induction*, also, those type-one speculations q such that q NC $(q')'$ only can be obtained by inductive reasoning.

In addition, and being p NC q, it is clear that no speculation q can be either a consequence ($p \leq q$), or a hypothesis ($q \leq p$) of p; hence conjectures are essentially of only four classes: consequences, hypotheses, type-one speculations, and type-two speculations. Of them, speculations are the only conjectures that were recently considered for the first time.

The following example can serve as an illustration of what is said. Suppose the information on the current atmospheric situation is:

It is midday, it is not sunny, and it is not an eclipse.

Denote by m = midday, e = eclipse, and s = sunny; then the premises are $p_1 = m, p_2 = s', p_3 = e'$, whose résumé is $p = (m \cdot s') \cdot e'$. Because $p \leq m \cdot s' \leq m$, and also $p \leq m \cdot s' \leq s'$, and $p \leq e'$, under transitivity it follows that m, s', and e', are obvious consequences, and s, e, and m' obvious refutations. Because it is not $(m \cdot s')' \leq p \leq m \cdot s'$, $(m \cdot s')'$ is not a hypothesis, and it should be a speculation. Because, for all x, it is always $p \leq p + x$, $(m \cdot s') \cdot e' + x$ is always a consequence, as are also $m + x$, $s' + x$, and $e' + x$, such as "It is midday or Monday," "It is not sunny or raining," and so on.

That a type-two speculation q is not directly linked with p and with p', does not mean to exclude the possibility of supposing some $r \leq q$ exists and is such that $p \leq r$, but keeping the incomparability (inferential isolation) of q with both p and p'. In a case like this, once it can be presumed that r is a hypothesis for q, and some q^* such that $r \leq q^*$ can be heuristically reached deductively from p and can substitute q in a "provisory form" could be, in some cases, a suitable approach to solving the problem.

It should be pointed out that, usually, deducing is not always an easy task; it suffices to think of some sophisticated mathematical proofs such as that of Fermat's last theorem done by Andrew Wiles. Deduction has in its favor the existence of deductive rules such as they are, for instance, those of the two basic modus ponens and modus tollens, able to secure the validity of deductive conclusions. In its turn, the processes of abduction can be conducted by backward deduction. Instead, speculating, and particularly in its type-two, is often based on rules of thumb, or on criteria not fully allowing the validity of the conclusions.

In deduction, the validity of the premises assures that of the consequences, but neither in abduction, nor in speculation, is the premises' validity sufficient for that of the conclusions. In this sense, and if correctly done, forward and backward deduction mean a kind of *safe* reasoning; the others are always *unsafe*. Note that hypotheses for p can be obtained by backward deduction, but that to select the better of them is a different problem.

8.4. Let's look at the property known as *monotony*, concerning the preservation of the conjecture's or refutation's character when the number of premises varies, and that is classically considered inferentially important to preserve, or not, the first-reached conclusions when the number of information items increases or decreases. As has been said, and usually, information flows more or less continuously.

For instance, if the number of three premises in $P = \{p_1, p_2, p_3\}$ increases up to four, $P^* = \{p_1, p_2, p_3, p_4\}$, with respective résumés p and p^*, and q is a consequence of P, the question is if q is also a consequence of P^*. Provided the above-mentioned properties of the inference relation $\leq$ were to hold, and under good conditions, from $p^* = ((p_1 \cdot p_2) \cdot p_3) \cdot p_4 \leq (p_1 \cdot p_2) \cdot p_3 = p$, and $p \leq q$, it would follow that $p^* \leq q$. Because this argument can be repeated, step by step, with more than three elements, it is concluded that the consequences of P are also consequences of P^*. Hence, consequences are monotonic; when information increases, the number of conclusions cannot decrease, and the former consequences are preserved; no one should be cancelled.

Also refutations are monotonic, because $p \leq q'$ and $p^* \leq p$ imply $p^* \leq q'$, under transitivity. Also the former refutations are preserved; no one should be cancelled.

Nevertheless, conjectures are antimonotonic: that is, if q is a conjecture of P^*, then it is also a conjecture of P: provided it were $p^* \leq p$, that would imply $p' \leq (p^*)'$, and also $q \leq / (p^*)'$, because if it were $q \leq p'$ it would be $q \leq (p^*)'$, which is absurd. It is possible that some former conjecture should be cancelled; it is not sure that all of them will be preserved.

Hence, conjectures being antimonotonic and consequences monotonic, the rest of the conjectures cannot in principle jointly enjoy the monotonic property, and, indeed, hypotheses and type-one speculations are antimonotonic.

For instance, if q is a hypothesis for P^* ($q \leq p^*$), from $p^* \leq p$, under transitivity it follows that $q \leq p$, and hypotheses are antimonotonic. If q is a type-one speculation from P^*, $q' \leq p^*$ implies (with $p^* \leq p$) $q' \leq p$, and being p NC q it should also be p^* NC q, because if it were $q \leq p^*$ then q would be a hypothesis for P^*, and if it were $p^* \leq q$ then q would be a consequence of P^*. There can be either former hypotheses, or former type-one speculations, that should be cancelled when the number of premises grows.

More premises imply no fewer refutations and no fewer consequences, but no more conjectures, no more hypotheses, and no more type-one speculations. More information leads to more possibilities for refuting and deducing, but to fewer possibilities when conjecturing for searching type-one speculations, and for abducing, all this, of course, under the above-presumed properties.

With respect to type-two speculations, examples with sets can show that, even in the setting of Boolean algebras, they are neither monotonic nor antimonotonic, and that hence they are properly *nonmonotonic*, are out of monotony and antimonotony; no law for their growing exists. This simply reinforces their wild inferential character as conjectures inferentially isolated from both p and p', and varying in a wild, neither monotonic nor antimonotonic, form. Very simple examples with precise statements (those represented by sets), even done with Venn diagrams, prove the nonmonotony of type-two speculations: it suffices to take, in a given universe, sets $P \subseteq P^*$, and finding sets Q_1 such that being type-two speculations for $P(Q_1^c \subseteq P \Leftrightarrow Q_1^c \cap P^c = \emptyset)$, are not so for P^*, and sets Q_2 that being type-two speculations for P^* are not so for P. Hence, the inductive reasoning for reaching type-two speculations certainly appears somehow separated from the rest of reasoning.

If consequences deploy what is already hidden in P, and hypotheses explain P, it rests for later on to clarify the "creative" role of speculations in reasoning; it is worthwhile remarking that there is a *kind* of reasoning leading to something new, or "creative reasoning," what laypeople often refer to as lucubrating or guessing, is inducing.

8.5. It is interesting to know when the presented model, even obviously naïve, is grounded on a solid basis, when Aristotle's principles of noncontradiction (NC), and excluded-middle (EM) can be proven to hold in it. In fact, and under a new interpretation of the Aristotelian word "impossible" as "self-contradictory," both principles can be submitted to proof. This is contrary to the usual idea, also coming from Aristotle, that they cannot be submitted to proof; they are just "principles" evident by themselves that cannot be reduced to simpler reasons.

Such an interpretation is:

$$p \text{ is impossible} \Leftrightarrow p \text{ is self-contradictory} \Leftrightarrow p \leq p'.$$

Hence for proving NC it should be proven that $p \cdot p'$ is always impossible, self-contradictory, that it is $p \cdot p' \leq (p \cdot p')'$, and, of course, taking care under which suppositions the proof can hold. Nevertheless, because Aristotle did not state EM so clearly, it is also proven after considering some definition for it.

Concerning NC, the proof is:

(a) $p \cdot p' \leq p$
(b) $p' \leq (p \cdot p')'$
(c) $p \cdot p' \leq p'$
(d) $p \cdot p' \leq (p \cdot p')'$, QED

by only adding the supposition that the triplet $(p \cdot p', p', (p \cdot p')')$ is $\leq$-transitive. Hence, and certainly, the "principle" can fail whenever a lack of transitivity can keep $(p \cdot p')'$ from $p \cdot p'$ inferentially isolated, or negation does not reverse the relation $\leq$, or the laws $p \cdot q \leq p$ and $p \cdot q \leq q$ fail.

Regarding EM, it seems that the closer form to what Aristotle stated of EM, is "$p + p'$ always holds", which can be understood as "not-$(p + p')$ is impossible", or "$(p + p')'$ is self-contradictory". That is, what should be proven is $(p+p')' \leq ((p+p')')'$, and the proof is:

(a) $p \leq p+p'$
(b) $p' \leq p+p'$
(c) $(p+p')' \leq p'$
(d) $(p+p')' \leq p+p'$, provided the triplet $((p+p')', p', p+p')$ were $\leq$-transitive
(e) $(p+p')' \leq ((p+p')')'$, QED

with the same former cautions on transitivity, the reversing character of negation, as well as the laws $p \leq p + q$ and $q \leq p + q$.

It should be pointed out that both NC and EM just correspond to ordinary deductive reasoning, and that, in particular, both principles should be preserved in formal deduction. They can fail by, for instance, the absence of transitivity.

Note that in a Boolean algebra the principles by self-contradiction collapse in a single "principle":

$$p \cdot p' \leq (p \cdot p')' \Leftrightarrow p \cdot p' = 0 \Leftrightarrow p + p' = 1 \Leftrightarrow 0 = (p+p')' \leq 1 = ((p+p')')',$$

thanks to holding the duality law $(p \cdot q)' = p' + q'$, equivalent to $(p+q)' = p' \cdot q'$, and the Boolean indistinguishability of contradiction, $p \leq q'$ with incompatibility $p \cdot q = 0$.

Obviously, in the general framework of ortholattices containing Boolean algebras, it also holds that $p \cdot p' \leq (p \cdot p')'$ is equivalent to $p \cdot p' = 0$. Hence what holds there is: $p \cdot p' \leq (p \cdot p')' \Leftrightarrow p \cdot p' = 0 \Leftrightarrow p + p' = 1$, and both principles are equivalent.

Nevertheless, in the framework of De Morgan algebras, also containing Boolean algebras, this chain of implications is not valid, and the principles only hold once stated by self-contradiction although, holding the duality laws, they can be seen as a single principle. In consequence, and in the general setting of ordinary reasoning, the two principles cannot be presumed as a single one.

Let's end this section with an example in the unit interval [0, 1] of the real line. With the conjunction ($\cdot$) given by the operation min, and the negation (′) by the function $1 - \mathrm{Id}$, it is obviously $p \cdot p' = \min(p, 1 - p) \leq p$, but were the conjunction given by the pondered mean, $p \cdot q = (p + 2q)/3$, it would be $p \cdot p' = (p + 2(1 - p))/3 = (2 - p)/3$, not always less than or equal to p, because $(2 - p)/3 \leq p \Leftrightarrow 2 - p \leq 3p \Leftrightarrow 1/2 \leq p$. Hence, in this case the NC principle fails.

Thus, the Aristotelian principles NC and EM strictly depend on which conjunction and disjunction should be respectively considered, and not only on transitivity that, in this example, is obviously valid.

8.6. The deductive chain, $p \leq q' \Rightarrow p \cdot q \leq q' \cdot q = 0$, valid in all ortholattices, shows that in this algebraic structure "contradiction" is "stronger" than incompatibility from, at least, the point of view of the inference relation $\leq$; there can be pairs of incompatible statements not being contradictory. In the setting of ortholattices, the equivalence between contradiction and incompatibility is bound to the particular and limit case of Boolean algebras that are almost characterized by the law of perfect repartition $p = p \cdot q + p \cdot q'$, for all pairs p, q. Notwithstanding, in the, even distributive, weaker structure of a De Morgan algebra, the former chain does not allow concluding $p \cdot q = 0$, because in them the law $q' \cdot q = 0$ does not hold; in this algebraic structure contradiction neither always implies incompatibility, nor reciprocality. For instance, the De Morgan algebra in [0, 1] given by the triplet (min, max, $1 - \mathrm{Id}$) is:

- $p \leq q' \Leftrightarrow p \leq 1 - q \Leftrightarrow p + q \leq 1$,
- $p \cdot q = 0 \Leftrightarrow \min(p, q) = 0 \Leftrightarrow p = 0, \quad \text{or} \quad q = 0$,

and it is incompatibility that implies contradiction, but not reciprocality; 0.3 and 0.6 are contradictory but not incompatible.

In general, both are independent concepts and, because with fuzzy sets their equivalence sometimes holds, sometimes their independence, it should be presumed that in plain language both concepts are, in general, independent, but without excluding that some implication between them can hold in a particular situation.

With the general relation $\leq$ of natural inference, usually not defined by $p \leq q \Leftrightarrow p \cdot q = p$ as it is in lattices, contradiction seems to be more natural than incompatibility; for instance, contradiction does not require counting with a (unique) null and minimum element that, in plain language, is not known how it should be defined, and that provided it were needed in some form should be artificially added to language; neither all parts of language, nor of mathematics, are ordered structures with minimum elements. In addition, it comes from experience that people not only can recognize contradiction, but hate it, usually try to escape from it, and stop reasoning when detecting contradictions. In this respect, it can suffice to remember the worries caused in "mathematical thinking" by the liar's paradox, the sorites paradox, and the like.

The definition $p \cdot q = 0$ comes directly from the classical set interpretation $\mathbf{P} \cap \mathbf{Q} = \emptyset$, from interpreting that statements p and q are incompatible if and only if the sets they respectively specify have an empty intersection and have nothing in common. It should be additionally remarked that provided $\mathbf{P} \cap \mathbf{Q}$ were reduced to a singleton, to contain a single common element, it would suffice to assert the "compatibility" of p and q; it seems not to exclude looking at incompatibility as a matter of degree. In plain language it is even sometimes presumed that p is compatible with q, provided p^* analogous to p, and q^* analogous to q, are known to be compatible.

In themselves, plain language and ordinary reasoning cannot be studied as purely algebraic structures; only some of parts of them admit to being represented by a strong mathematical model, and once the laws, or axioms, are checked, defining the model is fulfilled in the part of language under study.

References

1. E. Trillas, On 'crisp' reasoning with fuzzy sets. Int. J. Intell. Syst. **27**, 859–872 (2012)
2. E. Trillas, Glimpsing at guessing. Fuzzy Sets Syst. **281**, 32–43 (2015)
3. E. Trillas, On a model for the meaning of predicates, in *Views on Fuzzy Sets and Systems from Different Perspectives*, ed. by R. Seising (Springer, Berlin, 2009), pp. 175–205
4. E. Trillas, On the genesis of fuzzy sets. Agora **27**(1), 7–33 (2008)
5. G. De Cooman, Evaluation sets and mappings, the order-theoretic aspect of the meaning of properties, in *Introduction to the Basic Concepts of Fuzzy Set Theory and Some of its Applications*, ed. by E.E. Kerre (1991), pp. 159–213
6. K. Menger, A counterpart of Occam's Razor. Synthese **13**(4), 331–349 (1969)
7. G. Gentzen, Investigations into logical deduction, in *The Collected Works of Gerhard Gentzen* (North-Holland, New York, 1969)
8. C. McGinn, *Logical Properties* (Clarendon Press, Oxford, 2000)
9. E. Trillas, A. Pradera, A reflection on rationality, guessing, and measuring, in Proceedings of IPMU (2002), pp. 385–389
10. E. Trillas, S. Guadarrama, What about fuzzy logic's linguistic soundness? Fuzzy Sets Syst. **156**, 334–340 (2005)
11. E. Trillas, S. Cubillo, E. Castiñeira, On conjectures in orthocomplemented lattices. Artif. Intell. **117**, 255–275 (2000)
12. M. Ying, H. Wang, Lattice-theoretic models of conjectures, hypotheses and consequences. Artif. Intell. **139**, 253–267 (2002)
13. A. Fernández-Pineda, E. Trillas, C. Vaucheret, Additional comments on conjectures, hypotheses and consequences, in *Artificial Intelligence and Symbolic Computation*, eds. by J. Campbell et al. (Springer, Berlin, 2001), pp. 107–114
14. E. Trillas, A. Pradera, An orthogonally-based classification of conjectures. Mathware Soft Comput. **13**, 71–87 (2006)

15. E. Trillas, E. Castiñeira, S. Cubillo, Averaging premises. Mathware Soft Comput. **8**(2), 83–91 (2001)
16. A.R. de Soto, E. Trillas, A short note on counting conjectures. Mathware & Soft Comput. **14** (2), 165–175 (2007)
17. D. Qiu, A note on Trillas' CHC models. Artif. Intell. **171**, 239–254 (2007)
18. E. Trillas, I. García-Honrado, A. Pradera, Consequences and conjectures in preordered sets. Inf. Sci. **180**(19), 3573–3588 (2010)
19. I. García-Honrado, A.R. de Soto, E. Trillas, Some (unended) queries on conjecturing, in Proceedings of First World Conference on Soft Computing (2011), pp. 152–157
20. I. García-Honrado, E. Trillas, Characterizing the principles NC and EM in [0, 1]. Int. J. Uncertainty Fuzziness Knowl. Based Syst. **2**, 113–122 (2010)

Chapter 9
A Glance at Analogy

In beginning a study of analogy there are three immediate and basic questions:

(1) Which is the meaning of the word A = analogous (the binary mother-predicate of the analogy's concept), and how does it appear in reasoning?
(2) To what extent does analogy hold?
and
(3) In plain language, does analogy introduce a new type of conclusions?

Even without the possibility of presenting complete answers to such questions, they at least deserve to be posed.

9.1. For looking at a possible answer to the first part of the first question, we should recognize a relation $\leq_A$, "p is less analogous to q, than r is less analogous to s", linking pairs of linguistic labels (p, q) and (r, s) by taking into account the four relations $\leq_p$, $\leq_q$, $\leq_r$ and $\leq_s$, established in the respective universes of discourse: X for p and q, and Y for r and s. No such expression for $\leq_A$ is currently known, but provided it were described, and if mappings α from $X \times Y$ into [0, 1] can be found such that the number $\alpha(p, q)$ represents the extent to which p is analogous to q, and perhaps depending on a parameter reflecting that in which r is to s, then mappings α could be seen, although not necessarily as measures of "analogous," at least as indexes of analogy between p and q, provided p were to act in X, and q in Y. Examples of such indexes are the T-indistinguishability operators that are further considered for studying the breaking of synonymy chains.

It should be noted that in this view, and in principle, "analogous" does not admit saying "p is less analogous than q", because there are, at least, required intermediate terms r and s indicating to what p and q, respectively, are analogous.

As formerly said, for actually saying something interesting about a particular case of analogy, those indexes should be related to as many as possible characteristics that can be recognized in both p and q. For instance, concerning oranges and apples, by just taking into account the characteristic, or attribute, "spherical shape", it will be α (orange, apple) = 0.9, but if considering more attributes such as

E. Trillas, *On the Logos: A Naïve View on Ordinary Reasoning and Fuzzy Logic*,
Studies in Fuzziness and Soft Computing 354, DOI 10.1007/978-3-319-56053-3_9

color, taste, juice, and so on, then the number α (orange, apple) will be considerably lower than 0.9. Provided such requirement were satisfied, a suitable index's values could be of some help for controlling and avoiding the danger a superficial analogy can produce in reasoning such as that of confusing oranges and apples. By its own uncertain nature, analogy always should be controlled, and, when it is possible, a numerical control by means of some function α offers a possibility for appreciating its strength through its degrees.

Because neither the relation $\leq_A$ nor the specification of mappings α are easily found, this is a subject deserving of and waiting for deep analysis. Were it done, a scientific-like study of the, important by itself, subject of analogy could be ready to start; it is later reconsidered further. Nevertheless, it is an open problem, and especially, concerning the question of whether α can be a measure of the qualitative meaning $\leq_A$ of the word "analogous", for it and obviously $\leq_A$ should be previously expressed by means of the qualitative meanings of p, q, r, and s. For a deep study of the analogy's concept, it is essential to previously study the meanings its mother-predicate "analogous" shows in language.

9.2. Concerning the second part of the first question, and in a multitude of cases, it is considered that a linkage, or inference $p \leq q$, is analogous to another $r \leq s$, whenever it can be recognized that

$$\text{`}p \text{ is to } q, \text{as } r \text{ is to } s\text{', shortened by } p : q :: r : s[*].$$

where the crucial meanings of *is to* and *as* should be specified in each case. A possible way for dealing with such kinds of "qualitative proportionality", is to select a suitable mapping f with which $f(p) = r$ and $f(q) = s$ represent the analogy, that is, preserve the inference relation $\leq$ in the conditional form:

$$\text{If } p \leq q, \text{then } f(p) \leq f(q),$$

establishing the analogy between inferring $f(q)$ from $f(p)$ after inferring q from p, a typical inferential reasoning by analogy. For it, f must translate how the two *is to*, and the unique *as*, in [*], are understood, that is, which properties should be assigned to f for it. Were $\leq$ and f somehow specified, the linguistic equation in x:

$$p : q :: f(p) : x, [**],$$

perhaps could be solved through some calculus.

Of course, there are many different ways of understanding both the "is to" and the "as" in [*]. For instance, in ordinary life,

- Clouds are to rain, as gasoline is to fire.
- Book is to study, as blood is to injure.

are less precise than, in mathematics, are

- Three workers are to four hours, as five workers are to x hours.

– Segment A is to segment B, as segment C is to segment X.

of which those in the second case can be posed thanks to respectively knowing that there exist the direct arithmetic proportionalities,

– $3/4 = 5/x \Longleftrightarrow 3.\ x = 5.4 \Longleftrightarrow x = 20/3$.
– Length A/length B $\Longleftrightarrow$ length C/length $X \Longleftrightarrow$ length A. length X = length B. length C $\Longleftrightarrow$ length X = (length B. length C)/length A.

Were f given, it would follow that

$$f(3.x) = f(20),$$

and then, if f is bijective, the typical arithmetic solution $x = 20/3$ h would be obtained, and so on.

9.3. Provided, in relation to the second question, it were possible to associate numbers $t(p)$, $t(q)$, $t(f(p))$, and $t(x)$ to the terms in [**], and also translate the sign (:) by an operation (*) with numbers, a numerical equation would follow,

$$t(p) * t(q) = f(t(p) * t(x)),$$

from which, and if the properties of the operation * allow it, t (x) could be isolated in a form such as

$$t(x) = F(t(p) * t(q), t(f(p))),$$

with a two-argument function F. Note that t could be a measure of the statement's truth.

9.4. Concerning the third question, and whatever kind of function f can be for each p, it necessarily belongs to just one of the following three classes.

(1) It is $f(p) \leq p$; p can be forward inferred from its image $f(p)$, and it can be said that f is contractive when it happens for all p, or $f \leq$ Id. If $f(p)$ is not self-contradictory, then $f(p)$ is a hypothesis for p.
(2) It is $p \leq f(p)$; images $f(p)$ can be always be forward inferred from p, and it can be said that f is expansive when it happens for all p, or Id $\leq f$. If $f(p)$ is not self-contradictory, then $f(p)$ is a consequence of p.
(3) f is neither contractive, nor expansive at p, and in this case, $f(p)$ is isolated from p. Thus, and depending if it is additionally either $p \leq f(p)'$ or $p \not\leq f(p)'$, what are obtained are either refutations of p or conjectures of p that, in turn and provided it were also $f(p)' \leq p$, would be a type-one speculation, and if it is also $f(p)'$ isolated from p, would be a type-two speculation.

Note that f contractive will lead to hypotheses, f extensive to consequences, and that to reach speculations it is needed that, in similitude with the graphics of numerical functions, "f crosses the identity Id". With respect to refutations, it is enough that f makes $f(p) \leq p'$, that is, and in similitude with numerical functions, taking f below the function Id′ (e.g., below $1 - \text{id}$ in a numerical case in the interval [0, 1].

Hence, and through any analogy mapping f, the only statements $f(p)$ that can be reached from each p are consequences, hypotheses, speculations, or refutations. Independent of the additional properties f can enjoy, nothing different from refuting and conjecturing can be done by analogy under f, and, in this way, f-analogy appears as a "natural" way for just reaching refutations and conjectures.

Notwithstanding, the complete physical mechanism of reasoning, as a manifestation of the natural phenomenon "thinking", only will be fully explained and well comprehended after neurosciences can fully explain how thinking really works, what physical thought actually is. Only then will it be possible to know if analogy functions f really exist, and how they act if existing. Currently and for all this, there is still too great a lack of knowledge.

References

1. D. Hoffstater, E. Sander, *Surfaces and Essences: Analogy as the Fuel and Fire of Thinking* (Basic Books, New York, 2013)
2. A. Tversky, Features of similarity. *Psycho. Rev.* **84**(4), 327–352 (1972)
3. E. Trillas, J.M. Terricabras, A Scrutiny on representing meaning and reasoning. Archives for the Philosophy and History of Soft Computing, vol. 1 (2016), pp. 1–70
4. B. Bouchon-Meunier, L. Valverde, Analogy relations and inference; *Proc. FUZZ-IEEE*, pp. 1140–1144 (1993)
5. M. Rifqi, V. Berger, B. Bouchon-Meunier, Discrimination power of measures of comparison. Fuzzy Sets Syst. **110**(2), 189–196 (2000)
6. B. Bouchon-Meunier, M. Rifqi, S. Bothorel, Towards general measures of comparison of objects. Fuzzy Sets Syst. **84**(2), 143–153 (1996)
7. H. Farreny, H. Prade, About flexible matching and its use in analogical reasoning. *Proc. ECAI*, 43–47 (1982)
8. D. Gentner, Structure mapping: a theoretical framework for analogy. Cognit. Sci. **7**, 155–170 (1983)
9. S.J. Russell, *The use of knowledge in analogy and induction* (Pitman, London, 1989)
10. G. Gerla, Approximate similarities and Poincaré paradox. Notre Dame J Form. Logic **49**(2), 203–226 (2008)

Chapter 10
A Glance at Creative Reasoning

Once it has been shown that at least type-two speculations are born of the natural inference relation from the résumé and its negation, let's try to see how speculations are useful for "creating" what leads to either an explanation, or to find a consequence "hidden" in the résumé's premises. Guessing speculations often leads to introducing something new that, depending on its relevance, is or is not qualified as true creativity.

10.1. From examples such as

"He crashed against the wall and cried $\leq$ He cried",
and also
"He crashed against the wall and cried $\leq$ He crashed against the wall",

it can be agreed, as before, that the linguistic conjunction $(\cdot)$, enjoys the properties $p \cdot q \leq p$, $p \cdot q \leq q$. Hence, if p is the résumé of P, for any conjecture or refutation q, these two formulae hold.

Denoting a speculation by s, from $p \cdot s \leq p$, it is clear that $p \cdot s$ is, provided it were neither self-contradictory, nor inferentially equivalent to p, a hypothesis. Thus, "p and s" shows a way of obtaining hypotheses by conjunction between the résumé and a speculation from the set of premises $\{p, s\}$ and, provided the speculation were an inductive of type-one or type-two, reflects how hypotheses can be reached in an inductive or creative form.

Of course, that does not mean that all hypotheses are decomposable in the form of a conjunction of the résumé and a speculation (as happens in Boolean algebras), but just that some hypotheses are so; what is shown is that the set $\{p \cdot s\}$, for all speculations s, is just a subset of hypotheses for p.

Analogously, and from examples such as

"She bought book B" $\leq$ "She bought book B or book C",
and also
"She bought book C" $\leq$ "She bought book B or book C",

E. Trillas, *On the Logos: A Naïve View on Ordinary Reasoning and Fuzzy Logic*,
Studies in Fuzziness and Soft Computing 354, DOI 10.1007/978-3-319-56053-3_10

it can be agreed that the linguistic disjunction (+) enjoys the properties (also before supposed) $p \leq p + q$, and $q \leq p + q$, holding if p is the résumé of P, and q a speculation (s) from P. Thus, from $p \leq p + s$ it follows that $p + s$ ("p or s") is a consequence of P, showing a way to obtain consequences by disjunction between the résumé and a speculation, that is, from the set of premises $\{p, s\}$. Provided the speculation were of the inductive type-one or of type-two, it would reflect how consequences can be reached in a creative form.

Also this does not mean that all consequences are decomposable in such a disjunction, but only that some consequences are such; and in addition, the two results coincide with the experience researchers have about how some consequences, and hypotheses, are indirectly found.

Of course, the same two results hold when s is not a speculation from P but any non-self-contradictory statement. But it is nevertheless obvious that, even not deductively attainable from the résumé of P, any speculation $s = s(p)$ is related to P and, in this sense, both $p + s(p)$ and $p \cdot s\,(p)$ are not out of context involving the problem, but more linked with it than if, instead of $s(p)$, it were taken as a non-self-contradictory statement whatsoever.

These former (formal) results show that "speculating", is not just to confuse things; given the information supplied by the résumé, speculating from P can serve to explain P, and also to develop from it something hidden in the premises but not seen before speculating, in short, to obtain something that, in the relation of inference, is forward or backward p. Depending on the kind of speculation $s(p)$ is, the process of going from p to either $p \cdot s(p)$ (a hypothesis), or to $p + s(p)$ (a consequence), can be seen as less creative if $s(p)$ is of type-one, and more creative if $s(p)$ is of type-two. It recalls, in any case, the famous Archimedes of Syracuse shout, "*Eureka!*" "I got it!" usually considered a token of creativity. Conjecturing speculations, and mainly those of type-two, is the creative form of reasoning; it is the true "creative reasoning".

The actually undefined processes one can suppose existing, and by which a conjecture whatsoever is obtained, can be called "natural conjecturing processes", or guessing, and those for obtaining refutations, "natural refutation processes", or refuting. In particular, those for obtaining speculations can be called "natural speculation processes", those for obtaining ordinary consequences, "natural deduction processes", and those for obtaining hypotheses, "natural abduction processes".

Those unruled processes for reaching nondeductively reachable speculations are properly "inductive processes". Their formalization by, for instance, understanding what a "heuristics" really is, is still a pending subject that seems to be related to knowing something on an inductively searched conclusion. As said formerly and if, for instance, q is a type-two speculation of p and there is r such that $r \leq q$ and $p \leq r$, then supposing the triplet (p, r, q) is transitive, it would follow that $p \leq q$, and "speculating q from p", is transformed into "deducing q from p" through r; in this case, q can be reached algorithmically from p by two separate forward inferences. Of course, these processes require some additional and previous knowledge involving the goal q, such as that q "contains a part", r, allowing its backward linking with p. These are processes directed to a partially known goal.

The analysis of these processes in some suitable formal settings remains for a forthcoming study under the constraint of a mathematical structure, transitive and stronger than the very soft one currently presumed. For instance, it still cannot be seen how everything presented here can be reduced, in two limiting cases, to the soft structure of a basic fuzzy algebra, and to the strongest structure of a Boolean algebra, that is, the particular ortholattice structure on which classical logic is both posed and developed, as well as to intermediate structures such as De Morgan algebras and orthomodular lattices.

10.2. Notwithstanding, and in addition to its simplicity, the presented natural deduction-based model for laypeople's reasoning even allows the posing of some aspects of specialized reasoning such as the following on "falsifying" hypotheses.

It can sometimes be presumed that h is a hypothesis for P by, perhaps, some analogy with a similar known case, but without being able to "prove" if it is actually $h \leq p$ (p is forward inferable from h), or not. In these cases, there is a possible and slight turn known as the *deductive falsification* of h as a hypothesis for P. It comes from what follows, and only under the supposition that it is neither $h \leq h'$, nor is $h' \leq h$; that the singleton $\{h\}$ actually contains a premise.

- If q is a consequence of P, $p \leq q$, and it were $h \leq p$, under transitivity it would follow that $h \leq q$; that is, the consequences of P are also consequences of $\{h\}$. Hence, to falsify h deductively as a hypothesis for P it suffices to find a consequence of P that is not a consequence of h. This is the deductive way typically followed in mathematics to falsify a hypothesis.
- Provided the consequences of $\{h\}$ were conjectures of $\{h\}$, and because those of P are consequences of the presumed hypothesis h, for falsifying h it also suffices to find some consequence of P not being a conjecture of $\{h\}$. This is a not fully deductive way, typically followed in the experimental sciences to falsify a hypothesis inasmuch as, in these sciences, what is usually studied are conjectures of a presumed hypothesis.

Note that because all premises are also consequences, in both cases it suffices to find a single premise not being either a consequence or a conjecture of h to falsify the hypothesis h.

10.3. What has been presented is just a very simple mathematical model of the reasoning laypeople do, but without placing it in a strong formal framework as done, for instance, with the modeling of those kinds of specialized reasoning whose frameworks are those of Boolean algebras (classical reasoning), orthomodular lattices (quantum reasoning), or fuzzy algebras (imprecise reasoning) or, in a very particular case of it, De Morgan algebras (as in some approximate, or fuzzy reasoning). We just refer to the commonsense reasoning laypeople do in a plain language for dealing with daily life's usual decisions and for doing the corresponding actions. For instance, when a high school's last-year student chooses a university for after graduation, his or her decision is taken in an environment of imprecision and uncertainty that, without a formal setting's representation, is often done by analogy with some known examples; it analogously happens when a couple decides to be married, and so on. What is not taken into account is the "logically safe" character the reasoning's conclusions can show; it is considered further in a closer, although different, form as to what is commonly understood by logic, either deductive or inductive.

The material presented here is neither of a typically philosophical character, nor of a logical one; it can be perhaps said it is of a naturalistic, soft, and protologic character, inasmuch as it remains open to introducing constraints suitable enough for arriving at the strong models of mathematical logic devoted to several branches of specialized reasoning. It can be seen as a naïve although, perhaps, not properly a "kitchen" approach to natural laypeople's reasoning, in which analogy plays the fundamental roles of allowing the formation of concepts and facilitating first steps for speculating. Up to now, we have only tried to describe the possibility of establishing a kind of "naturalistic" mathematical representation of laypeople's ordinary reasoning independent of the validity of the conclusions that could be reached, of its logical safety.

It should be recalled that, contrary to the reasoning of the mathematical proof, ordinary reasoning often has unknown "jumps"; that is, in the reasoning's chains there can be consecutive pairs of statements that are supposed to be inferable from each other, but without knowing how it can actually be done, and as if it were the supposition that h is not a hypothesis for P without counting a falsification of h.

Of course, the concept of what is a heuristics way to reach a conclusion speculatively is not fully reported here, but will be, one hopes, studied at a forthcoming time. It should be noticed that the word "heuristics" refers, not necessarily, to what is done in, for instance, computer programs for playing chess where actually "heuristics" are but deductive chains followed after pruning the big tree of possible plays. In any case, those computational speculative heuristics ways consist in a deductive part for algorithmically reaching the conclusion. Algorithms are deductive, and necessary for the calculus computers can do.

A way for constraining the model is by representing the collectives, or usually imprecise concepts, by their shadows under the light of a concrete and contextual situation, sometimes purpose-driven, that is, by membership functions, states of fuzzy sets, or measures of the meaning of the predicates intervening in the statements once they are organized in elemental parts joined by connectives; as it is, in fact, the main idea behind Zadeh's new "computing with words." This is also considered at a forthcoming time within the very soft mathematical structure of basic fuzzy algebra (BAF) that allows considering imprecise predicates and does not impose strong properties (as in classical and quantum logics) neither to the representation of the linguistic connectives *and*, *or*, and *not*, as if they were the associative, the commutative, the distributive, the negation's strong character, and the duality laws; nor to those of the inference relation such as coming from a particular and pretended universal mode of representing conditional statements. These algebras allow representing plain language and ordinary reasoning in a more flexible form than Boolean algebras, orthomodular lattices, or De Morgan algebras can do by just identifying its structural ordering with the inference relation, and that, by the way, are particular cases of the symbolic basic abstract algebras that are introduced here later on.

Anyway, to reach a study of reasoning in a different way from that traditionally done in logic, it still lacks a way to capture directly the meaning of large statements without previously decomposing them in components joined by the connectives *and*, *or*, *not*, and so on, and capturing the respective meanings. As often happens, the situation is actually many times the reverse; first the full meaning of the phrase is captured, even in a gross form, and then the meanings of its components are captured. Possibly and in AI, it will be simultaneous with acquiring the capability of analyzing the meaning of large statements by computers, because they embody syntactic correctors, but currently lack semantic ones. It is a pending subject; without the capability of pragmatically understanding language's semantics, computers are unable to maintain large intelligent conversations. This is a goal for which fuzzy logic can help inasmuch as language is full of imprecise words.

10.4. Some comments are in order. Be it what it can be, what can follow from the mathematical constraints to which the presented naïve model could be submitted cannot be fully accepted before checking within language and reasoning. In any case, can it be expected to specify the suitable constraints by just purely mathematical thinking? Is it possible to conduct a mathematical analysis of reasoning without previously considering the nuances, flexibility, dynamism, and variability of the natural matter for such a study?

Plain language and ordinary reasoning are the necessary natural matters for it, thus what is actually needed is a new scientific methodology suitable to approach them, a natural science of language and reasoning; a kind of "physics" of reasoning that, instead of being grounded on matter and energy, is grounded on reasoning, with the use of language for it, and, doing it through the scientific processes typically consisting in systematic observation, controlled experimentation, and mathematical modeling. It should have something close to physics' mixing of experimental methodology and formal reasoning, where everything should be tested against the observed reality, and with the help of computer science, paraphrasing Eugene Wigner's words, allowing mathematics to show its "unreasonable effectiveness in the natural sciences." It should be pointed out that even a working mathematician trying to prove a conjecture formally, previously reasons the way people do by speculating on how he can proceed, based on his previous knowledge and, perhaps, by analogy with a former proof. For instance, Kurt Gödel proved his incompleteness first theorem, a recognized piece of mathematical creativity, by using the "diagonal method," before it was introduced for proving that the set of real numbers is not denumerable.

Such an open scientific-like modeling of reasoning and language, jointly with how the brain actually works, is one of the great challenges for science and technology in the twenty-first century. It could be a starting point for both the continuation of the Leibniz's hope expressed by his famous "*Calculemus!*" and the real possibility of building up machines' thinking like people do, something that, without a true knowledge of how plain language and ordinary reasoning actually work does not seem to be actually possible.

Along with such a new view for studying ordinary reasoning and natural language, their components of meaning, imprecision, the diverse types of uncertainty, and ambiguity, four linguistic phenomena that jointly with analogy permeate natural language and ordinary reasoning, seem to be important in the way towards undoing the so-called Gordian knot of artificial intelligence. Hence, the consideration of fuzzy sets (that meant, in 1965, the introduction of mathematical analysis for considering imprecision as did advocate John von Neumann) from the meaning's mathematical representation and measuring point of view deserves, in the immediate future, increasing attention to the modeling of natural language and ordinary reasoning. At least from this point of view, the understanding of fuzzy sets as (effectively measurable) quantities representing "meaning", can be seen as something relevant.

In any case, reasoning and meaning are strongly intermingled; both constitute a good part of rationality, although there are animals that think without having a language as people have; it seems clear enough that they cannot fully reason as humans do. The lack of linguistic labels for naming concepts is but a serious matter for reasoning; understanding the meaning of linguistic labels as a quantity can offer, indeed, abstract good aid in studying reasoning. Symbolic representation is essential for a scientific study of human reasoning.

Once seen that the predicates admitting a representation by membership functions are those whose primary meaning can be described by a graph, and that those P whose relation $\leq_P$ is empty are *meaningless,* those for which $\leq_P$ is not even imaginable could be called *currently metaphysical.* Historically, there have been many cases in which a metaphysical concept did acquire, later on, a meaning by sometimes either a change in the point of view from which they were considered, or from an advance produced by instruments for measuring, let's say, its intensity. It was, for instance, the case of passing to consider "irrational entities" such as the square root of two, to see them as "numbers" and to "create" the set of real numbers; it was a true manifestation of creativity allowing modern differential and integral calculus.

In some form, this classification of predicates concerns the struggle the members of the Vienna Circle underwent, in the first quarter of the twentieth century, against metaphysical concepts in philosophy, that is, from the grounds of the so-called analytic philosophy. Anyway, it should not be forgotten that there are metaphysical concepts that, carrying something else on their back, can be able to suggest analogies leading to new and useful measurable concepts. Nevertheless, it should be distinguished between suggesting ideas, and what can be safely established: if the first doesn't need to be measurable, the second requires it. To some extent, measurable predicates are empirically linked with the experience allowing them to describe something either physical or virtual.

10.5. Concerning nondeductive, inductive, or creative reasoning, their great importance in science, art, and philosophy, among others should be remarked, where new and fertile concepts are what actually guarantee their progress. In the words of Pablo Picasso, "What is creative is not transforming the sun in a yellow spot, but a yellow spot in the sun."

Creative speculation is not a type of reasoning properly considered in logic (almost always limited to analyze formally the rigid formal processes of deductive reasoning), once translated into an artificial language. Notwithstanding and before such specialization, it should be pointed out that "natural deduction", interpreted through the natural binary relation $\leq$ between statements, is the departing point for nonspecialized reasoning, once it is jointly taken with the concept of contradiction. In a, perhaps metaphorical, way it can be said that natural deduction is the mother of ordinary reasoning and that, to face specialized modes of reasoning, it should not be restricted to only formal deduction. Deduction alone is not sufficient for the progress of knowledge and, particularly, of scientific knowledge; it requires guessing or conjecturing; it requires creativity.

In this respect it is worthwhile quoting the words of the Nobel Laureate Sir Peter B. Medawar, "No process of logical reasoning can enlarge the informational content of the axioms and premises or observation statements from which it proceeds," implying that for extending the initial information to arrive at something new, logical deductive reasoning is insufficient. This is the right place for speculations and mainly those of type-two, the truly creative ones. Without induction, reasoning never would reach the successes which the history of science and technology are full of, but without its companion "formal deduction", both disciplines were not developed as they were after Galileo, or Newton, or Babbage, or Boole, or Einstein, or Gödel, or Turing, and so on. Deduction is basic for making knowledge solid; creation is essential for capturing what is still blowing in the wind.

10.6. With all that has been formerly said on the meaning of the word T = true, a window is open to study the predicate T when, being measurable, a group of people understands it in several forms, the importation of T to another universe of discourse when it is done by analogy, as well as that the collective T generates in X $[P]$ that, if P and T are both precise predicates, would reduce to the classical set of true statements. The true character of elemental statements can be translated into that of composed statements whenever the connectives, hedges, quantifiers, and so on, appearing in them are specified, and it is established how they behave with T. What is neither known, nor easy to see is, conversely, how a composed statement can be systematically obtained from the true character of its elemental parts.

Let's take into account the relationship between truth and inference, provided there were a measure t of "true" that can be applied to all statements under consideration. Suppose that t is nondecreasing for the inference relation $\leq$, and let $t(p) \in [0, 1]$ be the measure of the résumé p of the premises. If q is a consequence, from $p \leq q$ follows $t(p) \leq t(q)$; that is, the degree of *truth* of a consequence is, at least, that of the résumé. If h is a hypothesis, from $h < p$ follows $t(h) \leq t(p)$; that is, the truth of the hypothesis is, at most, that of the résumé. If r is a refutation, from

$p \leq r'$ follows $t(p) \leq t(r')$; that is, the truth of the negation of a refutation is, at least, that of the résumé. Note that if it were $t(r') = N(t(r)) = f^{-1}(1 - f(t(r)))$, with N a strong negation, then $f(t(p)) \leq 1 - f(t(r)) \Leftrightarrow t(r) \leq N(t(p)) = t(p')$ would follow; that is, the truth of a refutation is, at most, that of the negation of the résumé.

Finally, if s were a speculation its inferential isolation from p would not allow comparison of the true values $t(p)$ and $t(s)$. But, if s were of type-one, because it is $p' \leq s$, it would follow that $t(p') \leq t(s)$, and its truth would be, at least, that of the résumé's negation. Provided it were $t(p') = N(t(p))$, it would follow that $N(t(s)) \leq t(p)$. Nevertheless, if s were a creative speculation, its isolation from p' would not allow comparison of the truths of s, or s', with that of the résumé p.

Hence the truth of the negation of the résumé is an upper bound for the truth of refutations and a lower bound for type-one speculations, but nothing can be said for type-two speculations; they are also wild with respect to truth.

10.7. Moving towards a new experimental science of language and reasoning, it is relevant to count with the help of accurate measurements; there is no experimental science without controlled experimentation, and this often requires counting with mathematical models representing the basic system's variables with which some numerical, either real or complex, parameters can be computed to discriminate which one among the posed possible solutions is the best adapted to the known data.

What is being presented cannot be seen as a doctrine for modeling all that is in language. On the contrary, there are many aspects of both linguistic and logical character that up to now have not been modeled, and perhaps cannot be modeled by what has been shown. What is, notwithstanding, beyond doubt is that what has been presented actually enlarges the possibilities of representing plain language further than it does classical methods, but if it can be said that "language is all," it is very difficult to believe that a single mathematical model of "all language" can exist. For instance, language is full of ambiguity and no mathematical model of ambiguity is yet known, even if it can be imagined that, for predicates showing several meanings in the same context, their ambiguity could somehow be represented by modifying what has been presented for the meaning of imprecise predicates as a quantity. But if much of it is still waiting to be done, it seems clear enough that creative reasoning goes through analogy and speculation, through good guessing, and the breaking of transitivity.

10.8. As has been shown, most properties of the natural relation of inference ($\leq$) hold thanks to its presumed transitive law, without which there are serious doubts relative to the possibility of always keeping them. In the formalized forms of reasoning, the transitive law is often either taken for granted, or a direct consequence of its basic laws as they are, respectively, the cases in which the mathematical framework allowing translating the reasoning into a calculus is an ortholattice or a BAF. In them, transitivity cannot fail.

Nevertheless, transitivity cannot always be presumed when creative reasoning is done; it seems sometimes that there are triplets of statements (p, q, r), such that $p \leq q, q \leq r$, but $p \not\leq r$, cases in which a breaking of the inference chain does not allow us to reach r from p. Some aspects of creative reasoning seem to be closely related to the existence of "linguistic continua" in language, similar to the physical continuum with which Henri Poincaré identified the physical world in contraposition to the mathematical models describing it, and in which transitivity is always accepted. In this respect, two examples, first of a physical continuum, and second of a linguistic one, would be interesting.

Puncture point A in the hand's palm with a needle, and call a the corresponding sensation; repeat the puncture in a point B (with sensation b) at which b is not distinguished from a, and make a new puncture in a point C such that sensation c is not distinguished from b. As often recognized, c is clearly distinguished from a. It can be symbolically written, $a = b$, $b = c$, but $a \neq c$, revealing that skin's sensations do not constitute a transitive system, but Poincaré's physical continuum.

A typical linguistic continuum is constituted by a chain of words p, q, r, and so on, such that, in a dictionary of synonyms, it is recognized that q is a synonym of p, r is a synonym of q, s is of r, and so on. In this case it is always a word—let it be z for instance—such that it is not a synonym of p. Symbolically, indicating by $\leq$ that the dictionary allows us to infer the consequent as a synonym of the antecedent it is, $p \leq q, q \leq r, r \leq z$, but $p \leq\!/ z$. Semantic synonymy does not constitute a transitive system; it is, inside language, a linguistic continuum.

Because something can never be inferred from nothing, when it is not $p \leq q$, guessing q from p is a creative reasoning that requires having some knowledge of p, that is, some intellectual experience of what p means, its consequences and hypotheses, as well as how to refute it. Provided this collection of what is related to p were a linguistic continuum, q could only be obtained by a; on the contrary speculation, q can be obtained as either a consequence, or a hypothesis, or a refutation of p. Creative reasoning is related to the "discontinuities" between inferentially isolated concepts.

This is but an intuitive idea of what creative reasoning is, but it should still be added that there is a big difference between reasoning after knowing a persecuted and precise goal (for instance, when in mathematics the statement to be proven, the theorem, is read and its proof follows below it), and when the goal is unknown, or imprecise, and/or included in an environment of uncertainty, ambiguity, and the like as happens in a truly creative process of reasoning. In the first case, one is simply faced with capturing all the steps of a proof others made to answer the question of the theorem before it; something that is, nevertheless, of great importance for acquiring mathematical knowledge. In the second, one is faced with the intellectually exciting problem of stating a question only supported by the formerly reached conclusions, or previous information on the subject, often requiring being surrounded by doubts and experience with analogous situations. If questioning is always essential for creation, advancing knowledge needs to pose good questions and reach fertile answers.

For nonroutine research towards creating something new, one should be excited by doubts and questions. Creation requires it.

References

1. E. Trillas, A. Pradera, A. Alvarez, On the reducibility of hypotheses and consequences. Inf. Sci. **178**(23), 3957–3963 (2009)
2. E. Trillas, On 'crisp' reasoning with fuzzy sets. *Int. J. Intell. Syst.* (2012)
3. P.B. Medawar, *The Limits of Science* (Harper, London, 1984)
4. E. Trillas, Glimpsing at guessing. *Fuzzy Sets Syst.* (2015)
5. K.R. Popper, *Conjectures and Refutations* (Routledge and Kegan Paul, London, 1963)
6. R.A. Atchley, D.L. Strayer, P. Atchley, Creativity in the wild: improving creative reasoning through immersion in natural settings. *PLoSOne* **7**(12) (2012) (Journal on-line)
7. M.L. Ginsberg (ed.), *Readings in Nonmonotonic Reasoning* (Morgan Kaufmann Pubs, Los Altos, 1987)
8. E. Wigner, The unreasonable effectiveness of mathematics. Nat. Sci. Comments Pure Appl. Math. **13**(1) (1960)
9. J. von Neumann, The general and logical theory of automatas. in *Cerebral Mechanisms in Behavior*, ed. by L.A. Jeffreys (Wiley, New York, 1951)

Chapter 11
Formal Reasoning with Precise Words

Formal reasoning is a representation of either ordinary or specialized reasoning on some specific subject, provided the actual reasoning could be translated into a framework allowing a calculus for copying with it.

To do a formal reasoning is required, first of all and necessarily, counting with some mathematical framework where the reasoning could be translated into the calculus, and according to the reality existing behind it. The framework corresponds to the kind of reasoning to be translated into it, and allowing, as much as possible, its reproduction with the calculus; that is, from the characteristics the corresponding situation can show, and by fixing the basic properties or laws, the involved terms should verify when symbolically translated into the framework, that is, under the supposition that the chosen symbols and laws between them faithfully translate their meanings in the actual reasoning. In such a sense the framework should be as "natural", or suitable, as possible for each specific kind of reasoning.

Formal reasoning is, in the end, only a mathematical model of some particular specialized type of reasoning on something; hence, there is not exactly a single type of formal reasoning, but several mathematical models of it. At each specialized form of reasoning, it is supposed that the semantics of what is modeled is well translated into the corresponding mathematical model; for it, the internal laws of the representation's framework should be established according to what is recognized in the actual and external reasoning and its context. Sciences compact in artificial languages what, thought in plain language with scientific concepts, is considered basic for the corresponding subject and for formally developing the reasoning on it.

In what follows, the models for reasoning with precise words, and with both precise and imprecise words (and later in Part II, with the specialized reasoning physicists conduct on the quantum microworld), are considered. Basically, formal reasoning refers to mathematically formalized deductive reasoning, even if in the first two cases some hints regarding ordinary reasoning are presented. In addition, it should be noted that in these three cases, the corresponding inference relation is represented by a partial order verifying the transitive law; hence the former results requiring transitive triplets always hold. In addition, in these cases, the negation is

E. Trillas, *On the Logos: A Naïve View on Ordinary Reasoning and Fuzzy Logic*,
Studies in Fuzziness and Soft Computing 354, DOI 10.1007/978-3-319-56053-3_11

usually presumed to be strong, that is, verifying $(p')' \approx p$; thus also those results requiring one of the laws $p \leq (p')'$ or $(p')' \leq p$ also hold.

11.1. The mathematical framework credited as the undisputed one for classical reasoning with precise words is the theory of sets, that is, the structure of a Boolean algebra, as stated by Marshall Stone's characterization theorem of Boolean algebras in 1936. Such a framework comes from the "specification axiom" under which a precise word P acting in a universe of discourse X specifies a subset of X consisting in those x for which "x is P" holds, with its complement subset containing those x such that "x is P" fails; statements can only be either true or false.

In this case, the operations translating the linguistic conjunction $(\cdot)$, the disjunction $(+)$, and the negation $(')$, are supposed to verify all the laws of a Boolean algebra, that is, of a distributive lattice with a single strong negation. These laws affect all the statements composed by means of such connectives; the model presupposes that in the corresponding language all Boolean laws hold. In particular, it holds the law of perfect repartition $p = p \cdot q + p \cdot q'$, a law that jointly with the negation, the commutative, and associative laws of $+$, and the conjunction defined by duality, $p \cdot q = (p' + q')'$, characterizes Boolean algebras as Edward V. Huntington proved in 1933. Of course, the laws derived from those that characterize Boolean algebras also hold; for instance, the represented statements are presumed to verify $(p \cdot p) + (q \cdot p) = p + q \cdot p = p$, because $q \cdot p \leq p$; $(p \cdot q') \cdot (p' \cdot q) = (p \cdot p') \cdot (q' \cdot q) = 0 \cdot 0 = 0$; $p + p = p$, and so on.

In this case, if/then statements, the conditional ones, are supposed to coincide with "negation of antecedent, or consequent"; that is, "if p, then q" $(p \to q)$ is presumed to coincide with $p' + q$, that, provided the algebra were complete, in its turn coincides with Sup $\{z; p \cdot z \leq q\}$ as has been formerly shown. Hence, the truth values, $t(p \to q)$, are equal to $t(p' + q) = \max(1 - t(p), t(q))$, that equals 1 if and only if it is $t(p) = 0$, or $t(q) = 1$; the conditional only holds provided the consequent were to hold or the antecedent fails.

It is said that a linguistic statement p is a tautology when it is $p = 1$ (the maximum of the lattice); nevertheless, it can be statements q that, not being a tautology, have truth value equal to one, $t(q) = 1$. An if/then statement $p \to q$ represents a tautology if $p' + q = 1$ that, as was shown, is equivalent to $p \leq q$, the partial order of the Boolean lattice's part, defined by $p \cdot q = p$, or equivalently by $p + q = q$. For instance, $p + p' = (p \cdot p')'$ is a tautology, as well as are all linguistic statements whose translation into the algebra is represented by $p' + p \cdot q + p \cdot q' = p' + p \cdot (q + q') = p' + p = 1$, but are not a tautology those whose representation is $p + p \cdot q + p \cdot q'$ that are equal to $p + p = p$.

It should be pointed out that the idempotent laws of conjunction and disjunction, respectively, $p \cdot p = p$, and $p + p = p$, imply $t(p \cdot p) = F(t(p), t(p)) = t(p)$, and $t(p + p) = G(t(p), t(p)) = t(p)$, showing that $F = \min$, and $G = \max$, are, at least, suitable commutative and associative solutions of these equations, under which $t(p \cdot q) = 1 \Leftrightarrow t(p) = t(q) = 1$ and $t(p + q) = 1 \Leftrightarrow t(p) = 1$ or $t(q) = 1$.

Notwithstanding, in the case of lattices, and Boolean algebras in particular, it can be proven that with t ranging in $[0, 1]$, the only admissible pair (F, G) is (min, max);

where, for instance, with F = prod, and $p \cdot p = p$, from $t(p) = t(p) \cdot t(p)$ either $t(p) = 0$, or $t(p) = 1$ follows, and true or false statements will only exist, but not a single q with $0 < t(q) < 1$.

What about the basic point, in inference, with $p \rightarrow q = p' + q$? That is, what can be said when the set of premises is $P(\rightarrow) = \{p, p \rightarrow q\}$ whose résumé is $p \cdot (p \rightarrow q)$? This, of course, supposes that $p \cdot (p \rightarrow q)$ is not self-contradictory, which now simply means $p \cdot (p \rightarrow q) \neq 0$, because in Boolean algebras it is $r \leq s' \Leftrightarrow r \cdot s = 0$; thus, from $p \cdot (p \rightarrow q) = p \cdot (p' + q) = p \cdot q$, equivalent to $p \cdot q \neq 0$, it is not $p \leq q'$.

An element c is a consequence of $P(\rightarrow)$ provided it were $p \cdot (p \rightarrow q) \leq c$, and h is a hypothesis for $P(\rightarrow)$, provided $h \leq p \cdot (p \rightarrow q)$. Hence c deductively follows from $P(\rightarrow)$ provided $p \cdot (p \rightarrow q) \leq c$. The modus ponens inequality is obtained with $c = q$, that, as proven, is equivalent to $p \rightarrow q \leq p' + q$, showing that $p' + q$ is the greatest possible expression of the conditional, and that because it follows $p \cdot (p \rightarrow q) \leq p \cdot q$, $p \cdot q$ is also a consequence of $P(\rightarrow)$. Additionally, with the greatest conditional $p' + q$, c is a consequence of $P(\rightarrow)$ if and only if $p \cdot q \leq c$.

Concerning a hypothesis $h \neq 0$, that is, h is not self-contradictory, the inequality $h \leq p \cdot (p \rightarrow q) = p \cdot q$ shows that h should be a hypothesis for both p and q, and reciprocally because $h \leq p$ and $h \leq q$, imply $h = h \cdot h \leq p \cdot q$. In the case where $p \rightarrow q$ were not $p' + q$, the hypotheses for $P(\rightarrow)$ are just those $h \neq 0$ that are hypotheses of both p and $p \rightarrow q$.

What about refutations and speculations of $P(p' + q)$? Refutations r are characterized by $p \cdot q \leq r' \Leftrightarrow r \leq p' + q'$. Type-two speculations s cannot be characterized by any inequality, but only by s NC $p \cdot q$, and s NC$(p \cdot q)'$, or s NC$(p' + q')$. Those of type-one should verify s NC $p \cdot q$, and $s' \leq p \cdot q$, or $(p \cdot q)' = p' + q' \leq s$; they should be, on the order of the algebra, isolated from $p \cdot q$ but greater than $(p \cdot q)'$, and, in particular, simultaneously greater than p' and greater than q', because it is $p' \leq p' + q'$, and $q' \leq p' + q'$.

Of course, if there are always consequences and refutations it can happen that neither the hypotheses nor the speculations exist. In this respect, let's consider a simple example of some interest concerning the reasoning that, for making a bet, is done on the events that can appear in throwing a die.

The elemental directly observable events in the experiment are "appears one", "appears two", …, "appears six" points; hence, the universe of discourse can be taken to be $X = \{1, 2, 3, 4, 5, 6\}$, and all the possible events are its subsets, with the empty set Ø corresponding, for instance, to a failure in throwing the die. For instance, the event "appears odd points", corresponds to the subset $\{1, 3, 5\}$, the event "appears more than 3" corresponds to $\{4, 5, 6\}$, and so on. Note that the full set X corresponds to the "sure event", consisting in "obtaining any possible number of points", the only one at which no bet is allowed; X is the only premise for the reasoning.

Hence, because all subsets S of X verify $S \subseteq X$, and subset inclusion is the counterpart of $\leq$ in the power algebra 2^X of subsets, the events are but hypotheses; the bets are on hypotheses. The only consequence, $X \subseteq S$, is obviously X and the only refutation is the empty set because it is $X \subseteq S' \Leftrightarrow S \subseteq X' = \emptyset \Leftrightarrow S = \emptyset$, the one at which nobody will bet.

In this example, there are no speculations, a single consequence, a single refutation, and many hypotheses. The theory of probability mainly concerns the measuring of the chances hypotheses can have, the hypotheses that can be made on the possible results of a random experiment expressible by means of precise words. The case with either nonrandom experiments, or with "imprecise events", is considered later on.

Summing up,

- The uniqueness of the three operations translating the linguistic *and*, *or*, and *not*, as well as the great number of laws a Boolean algebra (or a power set) enjoys, makes the model a very simple one in which, for instance, refutations $r(p \leq r')$ coincide with those r such that $p \cdot r = 0$, that is, those subsets with empty intersection with the résumé's subset.
- Analogously, conjectures $p \leq / \, q'$ coincide with those q such that $p \cdot q \neq 0$, that is, those subsets with nonempty intersection with the résumé, and first-type speculations with those subsets s not comparable with the résumé but whose negation is included in it, $s' \leq p$, equivalent in this case to $p' \leq s$, and also to $p + s = 1$, because: $p' \leq s \Rightarrow 1 = p + s$, and $1 = p + s$ implies $p' = p' \cdot s \leq s$.

Hence, the Boolean model is uniform for all reasoning in which words are precise, all information on them is at least potentially available, and no degrees of truth beyond 1 and 0 are required. This kind of reasoning is typical of linguistic environments on which a perfect cut can be made between what is and what is not, where "ideal" perfect classifications can actually be obtained, something that is not always possible when the descriptions of situations or phenomena, either physical or virtual, are made with imprecise words, with uncertainty or with ambiguity, as is usual in ordinary reasoning and when, for instance, the behavior of a dynamical physical system is described in a plain language.

Nevertheless, there are descriptions of some situations that are done with precise words, as if they were some interesting random experiments such as that of throwing a die, and that are full of uncertainty; the events are describable in precise linguistic words, but are uncertain. In these experiments, the linguistic description of the events that can be obtained is well translated into a Boolean algebra of crisp sets, and, for computing the uncertainty of events, the idea of probability was first introduced, and later subjected to a very short, simple, and beautiful axiomatic, introduced by Anatoly N. Kolmogorov in 1933. This probability is, actually, a measure of the event's uncertainty when it is precisely describable, but is not the only interpretation of the probability's concept. The analysis of the meaning in language of the word "probable" is still open, the mother-predicate of the (abstract) concept of probability, which "measures of probability" are supposed to measure. In this interpretation, is probability a measure? And, what does it measure?

The answer is hidden in the same Kolmogorov definition, namely in the axiom of additivity:

If $p \cdot q = 0$ ($\Leftrightarrow p \leq q'$, p and q are contradictory), then prob($p + q$) = prob(p) + prob(q),

because in Boolean algebras it holds $p \leq q \Leftrightarrow q = p + q = p + p' \cdot q$, and it is $p \cdot (p' \cdot q) = (p \cdot p') \cdot q = 0 \cdot q = 0$, then prob($q$) = prob($p$) + prob($p' \cdot q$) $\geq$ prob(p). Thus, taking the lattice's order of the Boolean algebra as the qualitative meaning of "probable", the mapping prob, assigning numbers in [0, 1] to the events, is a measure of the word "probable". Kolmogorov's probability completes the graph $(2^X, \subseteq)$, identified with the qualitative meaning of probable in 2^X, to the triplets $(2^X, \subseteq, \text{prob})$ that, in this form, each can specify a full meaning of the word "probable". That is, Kolmogorov's probability is actually a measure of the linguistic qualitative meaning of probable, $\leq_{\text{probable}}$, under its identification with the relation $\subseteq$ of set's inclusion.

This is something that, perhaps acceptable with precise words, is not clearly so with the imprecise ones, whose meanings cannot be represented by crisp sets. Note that the supposition $\leq_{\text{probable}} = \subseteq$, comes from accepting that "less elements" can be identified with "less probable".

In conclusion, although the evident successes of the formal Kolmogorov theory of probability that, based on Boolean algebras, can lead to assigning great confidence in the former interpretation of the meaning of probable when used with precise words, it is still open to study when "probable" is used with imprecise words. Note that in plain language expressions such as, "It is with high probability that John is rich," in which neither "high" nor "rich" can always be constrained to be represented by crisp sets, are often uttered; examples like this cause us to look again at the meaning of the word "probable" in plain language.

11.2. In the classical case, with the inference relation identified with the lattice's order of the Boolean algebra, and because it is transitive, there is no room for inferential jumps in a deductive process, except an operative mistake or the ignorance of something that can produce an erroneous proof. In a correct proof there cannot be jumps; correct proofs are conducted in algorithmic form, that is, by enchaining statements in such a way that all steps in the chain hold thanks to its first step which is initially supposed to hold; each step is fired thanks to its former step, and by following the rule $p{:}p \leq q{::}q$, of modus ponens.

A proof of q from p consists in a sequence $\{p, p_1, p_2, \ldots, p_{n-1}, q\}$, such that $p \leq p_1$, $p_1 \leq p_2$, …, $p_{n-2} \leq p_{n-1}$, and $p_{n-1} \leq q$. Because $\leq$ is transitive, it follows that $p \leq q$; that is, q is deduced from p thanks to the inferential steps p_k regardless of its number. The chain $p \leq p_1 \leq \ldots \leq p_{n-1} \leq q$, is but an algorithm that allows reaching q from p; of course, it does not mean such an algorithm is unique, but usually several different proofs of q from p are available, and mathematicians prefer those with a minimum number of steps (often considered among the most beautiful proofs). As soon as the transitive law of $\leq$ is lost, algorithms can break at some intermediate step without allowing finally and safely concluding $p \leq q$, a proof of q from p.

Algorithms are essential for mechanizing formal deductive reasoning; from very early in the history of artificial intelligence, there have been computer programs or algorithms that proved some previously known mathematical theorems with fewer steps than proven by mathematicians. Let's remember the old case of the Herbert Simon program, Logical Theorist, that proved a theorem appearing in the book *Principia mathematica* by A.N. Whitehead and B. Russell, where it was proven with a larger proof than the one obtained by Logical Theorist. It was attained, nevertheless, in the short and closed context of the few axioms constituting the previous information needed for the proof.

11.3. As formerly observed, in the precise case, the NC and EM principles expressed in the former self-contradictory form collapse, respectively, in the equivalent Boolean axioms $p \cdot p' = 0$, and $p + p' = 1$, because, as is well known, in Boolean algebras: $x \leq x' \Leftrightarrow x = 0$, and $x' \leq x \Leftrightarrow x = 1$. Observe that because the reciprocal also holds: $p \cdot p' \leq (p \cdot p')' \Leftrightarrow p \cdot p' = 0$, and $(p + p')' \leq ((p + p')')' \Leftrightarrow p + p' = 1$.

Note that in non-Boolean ortholattices these equivalences also hold, but neither in De Morgan algebras nor in basic fuzzy algebras (BAFs) where, nevertheless, the principles only hold in their "self-contradictory form" provided the inference relation were taken coincidental with their respective orderings. In all these cases, the inference relation is transitive and there is no room for the failing of the principles. But in ordinary reasoning it cannot always be presumed that the inferential relation $\leq$ is an algebraic order, and less again is it always transitive. In ordinary reasoning, transitivity is a local property.

11.4. NC and EM can also be analyzed from the inferential point of view, and a hint on it follows. Note that in its former interpretation $(p \cdot p')'$ can be seen as simply being a refutation of $p \cdot p'$, and $((p + p')')'$ as being one of $(p + p')'$. Hence, they cannot be conjectures; but, what about $p \cdot p'$ and $p + p'$?

Provided p were not self-contradictory, taking the singleton $P = \{p\}$ as the set of premises, and presuming transitivity, $p \cdot p'$ cannot be a consequence of P, $p \leq p \cdot p'$, nor p a hypothesis for $p \cdot p'$. Were it a consequence, and the triplet $(p, p \cdot p', p')$ transitive, because it is $p \cdot p' \leq p'$, it would follow that $p \approx p \cdot p'$, and because of NC, the absurd $p \leq p'$ would also follow. In addition, the possibility that $p \cdot p'$ is a speculation can be avoided inasmuch as $p \cdot p' \leq p$ implies that it is not p NC $p \cdot p'$. Statement $p \cdot p'$ is not a conjecture of $\{p\}$ and, hence, should be a refutation of $\{p\}$, and it is so because $p \cdot p' \leq p'$. But, on which conditions can it be $p \leq (p \cdot p')'$? Because $p \cdot p' \leq p'$, it is $(p')' \leq (p \cdot p')'$, it suffices to count with $p \leq (p')'$, and the transitivity of the triplet $(p, (p')', (p \cdot p')')$ to have $p \leq (p \cdot p')'$.

On the contrary, because it is always $p \leq p + p'$, $p + p'$ is a consequence of P, and p a hypothesis for $p + p'$.

References

1. E. Trillas, I. García-Honrado, Hacia un replanteamiento del cálculo proposicional clásico. Agora **32**(1), 7–25 (2013)
2. M.H. Stone, The theory of representations of boolean algebras. *Trans. AMS* **40**, 37–111 (1936)
3. P.R. Halmos, *Naive Set Theory* (Van Nostrand, New York, 1960)
4. E.V. Huntington, Sets of independent postulates for the algebra of logic. Trans. AMS **5**, 288–309 (1904)
5. E. Trillas, On functions that cannot be mv-truth values in algebraic structures. Stochastica **12**(2-3), 223–227 (1988)
6. A.N. Kolmopgorov, *Foundations of the Theory of Probability* (Chelsea, New York, 1956)
7. D.A. Kappos, *Probabilistic Algebras and Stochastic Spaces* (Academic Press, New York, 1969)
8. H.A. Simon, *Models of My Life* (The MIT Press, Cambridge, 1996)
9. J.-P. Dubucs, *Philosophy of Probability* (Kluwer, Dordrecht, 1993)
10. E. Trillas, I. García-Honrado, S. Termini, Some algebraic clues towards a syntactic view on the principles of non-contradiction and excluded-middle. Int. J. Gener. Syst. **43**(2), 162–171 (2014)
11. A.J. Ayer, *Probability and Evidence* (Columbia University Press, New York, 1972)
12. S. Rudeanu, *Axioms for Lattices and Boolean Algebras* (World Scientific, Singapore, 2008)

References

1. [illegible]
2. [illegible]
3. [illegible]
4. [illegible]
5. [illegible]
6. [illegible]
7. [illegible]
8. [illegible]
9. [illegible]
10. [illegible]

Chapter 12
Formal Reasoning with Imprecise (and Precise) Words

Without changing its meaning, imprecise words cannot be represented by crisp sets, but by the measures or the membership functions of fuzzy sets whose linguistic labels are those imprecise words. Representing the reasoning with imprecise words, for reaching Zadeh's computing with words, a calculus with fuzzy sets is necessary, and for which BAFs (basic fuzzy algebras) only facilitate a too general skeleton. To such a computing goal, BAFs need to be specialized to each particular problem, by defining operative forms for conjunction, disjunction, and negation, as well as by finding suitable expressions for conditional statements, modifiers, and quantifiers. One method is to suppose that such operations are functionally expressible by functions/operations endowed with laws that, without allowing the Boolean algebra structure, could permit some computations once chosen, as accordingly as possible, to that which the designer considers is in the corresponding setting. Among such operations, the most widely used are continuous t-norms, continuous t-conorms, and strong negations. Because negations were considered before, let's briefly consider the first two in which "*t*" comes from its former introduction by Karl Menger, and refers to "triangular", in as much as they were first used for establishing a triangular inequality in probabilistic metric spaces, where distances are not numbers but probability distributions. With these functions, the standard algebras of fuzzy sets can be defined and the study of their laws can begin.

12.1. A continuous t-norm T is a continuous, commutative, monotonic, and associative binary operation in [0, 1] such that has 1 is neutral, 0 is absorbent, and is nondecreasing in its two variables; *t*-conorms S are operations in [0, 1] obtained from t-norms T in the form $S(x, y) = 1 - T(1 - x, 1 - y)$. They are commutative, associative, and nondecreasing in both variables, with 0 neutral and 1 absorbent; It is obvious that T is continuous $\Leftrightarrow$ S is continuous.

Note that $y \leq 1$ implies $T(x, y) \leq T(x, 1) = x$, and $T(x, y) \leq T(1, y) = y$, thus $T(x, y) \leq \min(x, y)$; that is, the greatest t-norm is $T = \min$, a continuous one. Analogously, it is $\max \leq S$ for any t-conorm S; max is the smallest t-conorm and it is continuous. Hence, for any pair whatsoever of a t-norm T, and a t-conorm S, it is $T \leq \min \leq \max \leq S$; in particular, it is always $T \leq S$ and $T =/S$. In addition to

E. Trillas, *On the Logos: A Naïve View on Ordinary Reasoning and Fuzzy Logic*, Studies in Fuzziness and Soft Computing 354, DOI 10.1007/978-3-319-56053-3_12

min, two typical instances of continuous t-norms are $T(x, y) = \text{prod}(x, y) = x \cdot y$, and $T(x, y) = W(x, y) = \max(0, x + y - 1)$.

Note that the definition of a t-conorm S shows its "duality" with a t-norm T with respect to the strong negation $1 - Id$, but that, analogously, t-conorms S can be defined by duality with t-norms T by any strong negation N: $S(x, y) = N(T(N(x), N(y))$, for all x, y in [0, 1].

Additionally, it should be pointed out that, contrary to the continuous upper bound min for all t-norms, there is no continuous lower bound for them but that, as is easy to prove, such a bound is given by the discontinuous t-norm

$$Z(x, y) = \min(x, y), \text{if } x = 1, \text{or } y = 1, \text{and } Z(x, y) = 0, \text{otherwise},$$

and obviously verifying $Z \leq T$ for any t-norm T; Z is the lower bound for all t-norms, and not only for the continuous ones. In the same vein, although max is the continuous lower bound of all t-conorms, the upper bound of all t-conorms is the discontinuous t-conorm

$$Z^*(x, y) = 1 - Z(1 - x, 1 - y) = \max(x, y), \text{if } x = 0, \text{ or } y = 0, \text{ and } Z^*(x, y) = 1, \text{otherwise},$$

the dual t-conorm of Z with respect to the negation $1 - Id$.

Hence, all t-norms T verify $Z \leq T \leq \min$, and all t-conorms S verify $\max \leq S \leq Z^*$; all of them are contained, respectively, in the functional closed intervals [Z, min] and [max, Z^*]. Hence, for all pairs (T, S), not only continuous, $Z \leq T \leq \min \leq \max \leq S \leq Z^*$, although neither all the functions in [Z, min] are t-norms nor are all those in [max, Z^*] t-conorms. There exist an enormous amount of t-norms and t-conorms, of which only the continuous are fully characterized.

Given a t-norm T, for each order automorphism f on the ordered unit interval, the function $T_f(x, y) = f^{-1}(T(f(x), f(y))$ is also a t-norm that is continuous if and only if T is. Analogously, for a t-conorm S, the function $S_f(x, y) = f^{-1}(S\,(f(x), f(y))$ is also a t-conorm that is continuous if and only if S is. It is said that all the t-norms T_f constitute the family of T, and all the t-conorms S_f the family of S, of course, taking into account all the order-automorphisms f of the unit interval. For instance, the family of min is reduced to min, because $f^{-1}(\min(f(x), f(y)) = \min(x, y)$; the family of prod are the continuous t-norms $\text{prod}_f(x, y) = f^{-1}(f(x) \cdot f(y))$; and the family of W are the continuous t-norms $W_f(x, y) = f^{-1}(W(f(x), f(y))$.

Note that an automorphism of the ordered unit interval ([0, 1], $\leq$) is a mapping f: [0, 1] $\rightarrow$ [0, 1] that is strictly nondecreasing, and verifies $f(0) = 0, f(1) = 1$; thus, all automorphisms are continuous functions.

Any t-norm verifies $T(x, x) \leq x$, for all x, but the only one verifying $T(x, x) = x$, for all x in [0, 1] is $T = \min$, even if it does not mean that, for some continuous t-norm T, no point x different from 0 and 1 can verify $T(x, x) = x$(T is idempotent at point x). This is not the case of those t-norms in the former two families: prod $_f(x, x) = f^{-1}(f(x) \cdot f(x)) = x \Leftrightarrow x \in \{0, 1\}$, and $W_f(x, x) = f^{-1}(\max(0, f(x) + f(x) - 1)) = x \Leftrightarrow \max(0, 2f(x) - 1) = f(x) \Leftrightarrow f(x) = 0$, or $f(x) = 1 \Leftrightarrow x \in \{0, 1\}$.

Hence, continuous t-norms are min (with all x idempotent); those without idempotents other than 0 and 1, and those with some idempotent elements different from 0 and 1, of which the second are just those in the families of prod and W, and the third are called "ordinal sums" of t-norms.

All that is analogous with continuous t-conorms: there are only max, the family of $1 - \text{prod}(1 - x, 1 - y) = x + y - x.\, y$, denoted either by Sum-prod, or prod*, the family of $1 - W(1 - x, 1 - y) = \min(1, x + y)$, denoted by W^*, and called "bounded sum", and the ordinal sums of t-conorms.

12.2. The classification of continuous t-norms, and t-conorms, helps to compute with them and also to obtain theorems concerning the verification of some laws through solving the corresponding functional equations or inequalities, for instance, to know if, and when, the typical Boolean law of perfect repartition can hold with imprecise statements, that is, for studying the validity of the equation $\mu = \mu \cdot + \mu \cdot \sigma'$ for all μ and σ in $[0, 1]^X$. This can be known, under the hypotheses that conjunction, disjunction, and negation are functionally expressible by, respectively, a continuous t-norm T, a continuous t-conorm S, and a strong negation N, by solving the functional equation.

$$a = S(T(a,b), T(a,N(b)), \text{for all } a, b \text{ in}[0,1],$$

among whose solutions is the triplet $T = \text{prod}$, $S = W^*(=\min(1, \text{sum}))$, and $N = 1 - id$:

$$W^*(x \cdot y, x \cdot (1-y)) = W^*(x \cdot y, x - x \cdot y) = \min(1, x \cdot y + x - x \cdot y)$$
$$= \min(1,x) = x,$$

showing that such law holds in algebras of fuzzy sets in which no duality holds, something that has no place in the reasoning with precise words where the laws of duality are accepted to hold universally. Hence, in the imprecise case the law of perfect repartition can hold, and all the positive cases with continuous t-norms, t-conorms, and strong negation can be found by solving the former functional equation; these solutions are $T = \text{prod}_f$, $S = W_f^*$, and $N = N_f$, and hence the law of perfect repartition with imprecise words is not compatible with duality. This occurs, at least, in the BAFs that are standard algebras, that is, functionally expressible by continuous t-norms, continuous t-conorms, and strong negations; but it suffices for asserting that in plain language such a law cannot universally hold. Note that in the algebra expressed by the triplet (prod, W^*, $1 - id$), it holds the law of excluded-middle, because $W^*(x, 1 - x) = \min(1, x + 1 - x) = 1$, but not the law of contradiction, because $\text{prod}(x, 1 - x) = x(1 - x) = 0 \Leftrightarrow x = 0$, or $x = 1$, two laws also undisputed with precise words.

More laws of the classical crisp logical calculus can hold in the imprecise case and under diverse algebras, but always at the cost of losing some other Boolean laws. As was reported at the beginning of the former section, there is no single calculus able to formalize the reasoning with imprecise words, and an open question remains of finding the diverse possible ones that can be developed by means of several algebras of fuzzy sets, something actually important for seriously facing computing with words.

The limiting of the calculus with fuzzy sets to continuous operations occurs because most membership functions of fuzzy sets with an imprecise linguistic label are continuous as a manifestation of their flexible character and thus, provided the operations were not continuous, some unwished discontinuities would be added, in a perhaps artificial form, to the final computations. In addition, provided the membership functions showed some discontinuity, the continuous operations surely would not modify them. Usually, in the applications, membership functions are continuous functions $R^n \to [0, 1]$, with the discontinuous ones just referring to crisp sets. It is worthwhile remembering why the membership functions of the imprecise use of "small" in [0, 1] should be taken as continuous; if x were small, all points in a short interval $(x - \varepsilon, x + \varepsilon)$ would also be small.

12.3. What about representing the imprecise conditional statements "If μ, then σ" with the membership functions μ and σ specifying imprecise words? In this case, the rules of modus ponens (MP) and modus tollens (MT) cannot always be posed as they are in the classical crisp case because most of the systems that are describable by means of imprecise rules are not static, but dynamical.

When observing the reality of a rule "If μ, then σ", what is actually observed is not exactly μ, but μ^* "close to μ", or σ^* "close to σ". Hence, such rules of inference should be posed in the forms:

- 'If μ, then σ; μ^*: σ^*', with input μ^*, and output σ^*, for modus ponens,

and

- 'If μ, then σ; not σ^*: not μ^*', with input not σ^*, and output not μ^*, for modus tollens,

representing such approximate modes of imprecise reasoning with the constraint that $\mu^* = \mu$ should imply $\sigma^* = \sigma$ for preserving the classical case that, in addition, could really appear.

The corresponding inequalities within a standard algebra, a functionally expressible BAF by continuous t-norms, t-conorms, and strong negations, are:

$$T(\mu * (x), J(\mu(x), \sigma(y))) \leq \sigma * (y), \; for \; MP,$$

and

$$T(N(\sigma * (y)), J(\mu(x), \sigma(y))) \leq N(\mu * (x)), \text{ for MT},$$

for which verification the functions T, J, and N, should be designed in each context once the membership functions are, and by taking care that with $\mu^* = \mu$, $\sigma^* = \sigma$ should hold. Both inequalities are obviously equivalent to, respectively,

$$\text{Sup}_{x \in X} T(\mu * (x), J(\mu(x), \sigma(y))) \leq \sigma * (y), \text{for all } y \text{ in } Y,$$

and

$$\text{Sup}_{y \in Y} T(N(\sigma * (y)), J(\mu(x), \sigma(y))) \leq N(\mu * (x)), \text{for all } x \text{ in } X.$$

In this way, the variable affecting the input is eliminated at each of the left members of these inequalities, and they allow computing the (theoretical) outputs σ^*, and $N \circ \mu^*$ by "defining" them as, respectively, the left members in the two former inequalities. These rules for computing approximate imprecise outputs, introduced by Zadeh, are known as the compositional rules of inference (CRI); at least the first showed its usefulness in many applications of fuzzy control, where the function J representing the conditional is just taken to be min or prod.

Let's show examples with these two functions in $X = Y = [0, 1]$, with $\mu(x) = x$, $\sigma(x) = 1 - x/10$, $\mu^*(x) = 1$, if $x \in [0.7, 1]$, and 0 otherwise, and $\sigma^*(x) = 1$, if x [0, 0.2], and 0 otherwise.

(a) With $J(x, y) = \min(x, y)$, it is:
$\text{Sup}_{x \in [0,1]} \min(\mu^*(x), \min(x/10, 1 - y/10) = \min(1/10, 1 - y/10)$, and it follows that the theoretical output σ^* is $\sigma^*(y) = \min(1/10, 1 - y/10)$, the truncation of the σ's graphic by 1/10. Because a linguistic label for σ is "small", one for σ^* could be, for instance, "almost small" even if such name's attribution is to be checked with the given problem data.
(b) With $J(x, y) = x. y$, it is:
$\text{Sup}_{y \in [0,1]} \min(1 - \sigma^*(y), x/10 \cdot (1 - x/10)) = \min(1, x/10 \cdot (1 - x/10)) = x/10 \cdot (1 - x/10)$, and the theoretical output μ^* is the function $\mu^*(x) = 1 - x/10 \cdot (1 - x/10)$. Because a linguistic label for "not μ" is "not big", one for μ^* could be "not (big and small)", and whose suitability, as in the former case, should be checked by the designer.

Observe how the final linguistic naming of the outputs depends on the chosen algebra for computing; this shows the care that should be taken for choosing it in a design process. Note also that provided it were possible to take $T = \min$, and because this function is the greatest t-norm, the theoretical outputs would then be closer than possible (with continuous t-norms) to the theoretical outputs σ^* and $N \circ \mu^*$.

Let's repeat that in the case of imprecise reasoning a universal algebra for computing with fuzzy representations does not exist; at each practical problem, or application, the algebra should be contextually and carefully chosen according to the data, and once that has all been submitted to scrutiny by the designer. This is akin to constructing a house: it cannot be safely done without previously studying the ground, the house's form, its sustainable structure, the materials to be employed, and so on.

Let's repeat that the MP inequalities for those functions J hold with the biggest t-norm min because, for any previously chosen t-norm T, it is $\min(a, T(a, b)) \leq T(a, b) \leq b$. Because it is obvious that the former MP inequalities also hold for any function F such that $F \leq \min$, and, in particular for any continuous t-norm, making it equal to min the "defined outputs" become the bigger ones and, presumably, closer to the "real outputs" than with any other t-norm; this is why $T = \min$ is always taken in the CRI in fuzzy control.

Among the conjunctive representation of rules/conditionals "If μ, then σ", in the form $\mu \rightarrow \sigma = \mu \cdot \sigma$, $T = W$ is never taken but only $T = \min$, or $T = \text{prod}$ in $J(x, y) = T(x, y)$. Why? Because the t-norm W and all the t-norms in its family have zero-divisors, that is, numbers $x > 0$, $y > 0$, such that $W(x, y) = 0$. Namely it is $\max(0, x + y - 1) = 0 \Leftrightarrow x + y \leq 1$ and, for instance, $W(0.6, 0.4) = 0$; hence, a rule whose antecedent and consequent hold with a positive degree could be not fired for having a joint zero degree. Of course, neither min nor prod has zero-divisors because they are $\min(x, y) = x \cdot y = 0 \Leftrightarrow x = 0$, or $y = 0$. Additionally, only prod is strictly nondecreasing in both variables, but min and W are not; for instance, it is $0.4 < 0.5$ and $0.3 < 0.5$, and $0.4.\ 0.3 = 0.12 < 0.25 = 0.5.\ 0.5$; but $W(0.4, 0.3) = W(0.5, 05) = 0$, and, with $0.6 < 0.7$, it follows that $\min(0.4, 0.6) = \min(0.4, 0.7)$.

12.4. Representing systems described by several imprecise linguistic rules with the same function J for all the rules deserves comment. The reasonability of such election depends on a uniform meaning of the linguistic conditionals, or rules, and on the suitability of a same proto-form for all the rules; were such the case, there is nothing against it. Nevertheless, language comprises conditional statements more complex than the rules of these systems are. For instance, "If (If p, then q), then (If not p, then r)" in classical calculus is expressed by $(p' + q) \rightarrow (p + r) = (p' + q)' + (p + r) = p \cdot q' + p + r = p + r = (p')' + r = p' \rightarrow r$; that is, it is equivalent to "p or r", or to "If not p, then r". Anyway, provided the intermediate conditionals were to be expressed in conjunctive form, the classical representation would be $p \cdot q \rightarrow p' \cdot r = p' + q' + p' \cdot r = p' + q'$, whose meaning is that of "not p or not q"; but, provided the three conditionals were conjunctive, then it would be $(p \cdot q) \cdot (p' \cdot r) = (p \cdot p') \cdot q \cdot r = 0$, and the statement would be meaningless. The meaning of $p \rightarrow q$ depends on its interpretation.

Hence, in the case of a reasoning consisting in several conditionals with precise words, the meaning of each conditional affects the meaning of the full statement; it must carefully proceed in its representation, and even more with imprecise words. For instance, in a fuzzy case with two rules $\mu \rightarrow \sigma$, and $\alpha \rightarrow \beta$, functionally represented by $J_1(\mu(x), \sigma(y))$, and $J_2(\alpha(x), \beta(y))$, the forms J_1 and J_2 can show should be previously studied before taking $J_1 = J_2$; that is, before supposing that the meanings of the two conditionals made them coincide in the same proto-form. This is something not usually done in fuzzy control where, not with standing, the usual simplicity and understandability of the rules does not make taking different functions J strictly necessary. But this is not the case in parts of language where the conditionals can be neither simple nor easily understandable.

In a novel by the Spanish writer Enrique Vila-Matas, the complex statement appears, here translated from Spanish into English,

> "I had always told myself that if life has no sense neither has reading, but suddenly it seemed to me that the process of reading and searching for artists of the Not, did have a lot of sense. Unexpectedly, I felt that the search for bartlebys gave sense to my life."

This statement hides a reasoning involving three conditionals:

If (If life has no sense, neither has reading), then (If reading has sense, also has life), and that can be symbolically represented by the formula $(p \to q) \to (q' \to p')$. In the classical mode of reasoning, and provided everything in it were precise, it is $(p \to q) \to (q' \to p') = (p' + q)' + (q + p') = (p' + q)' + (p' + q) = 1$. But, were the conditional expressed in conjunctive form, this formula would be a contradiction, because $(p \cdot q) \cdot (q' \cdot p') = (p \cdot p') \cdot (q \cdot q') = 0$.

Nevertheless,

(a) Sense is taken in the text as an imprecise word, inasmuch as it is said in it that reading has "a lot of sense."
(b) The author does not identify contrasymmetrical conditionals, because he separates, "If life has no sense, neither has reading," from, "If reading has sense, also has life."
(c) Contextual information on the author's literary biography states that he is a passionate reader writing in a dense and complex literary style.

Hence it does not seem that the truth of the statement could be considered out of context with precise words and representing conditionals in a unique form containing the property $p \to q = q' \to p'$. Its contextual truth deserves to be analyzed by representing in fuzzy terms all that is involved in the statement; that is, through a semantic analysis of the statement.

Once this is done, the result is that the only admissible function J is J_{prod}, and the truth value of the triple conditional is 1 between the values $1 - t$ and 1, with t the degree of "reading has sense" supposedly very high for Vila-Matas (hence 1 − t is very small), and between 0 and $1 - t$ is a nondecreasing curve. Hence, the triple conditional is almost true, with the zone $[0, 1 - t)$ in which it can be not totally true, and that is less significant as t is greater. For instance, were $t = 0.91$ the "dangerous" zone for truth would be reduced to $[0, 0.09)$. Vila-Matas' triple conditional can be considered as being almost always contextually true except if t is not big, something that the information on Vila-Matas seems to avoid.

Note that the former is but a semantic analysis, done "by hand" of an imprecise statement. Is it possible to do it by computer? Provided it were someday possible, the computer would count with an automatic "semantic analyzer" not only able to perform analysis like the former, but to send the writer questions such as "Do you believe 'reading has sense' is with a high degree?" and, in the supposition the author answers "not", or "not always", "not necessarily", or "I don't know", and so on, continues asking him up to a final recommendation such as "Be careful. The statement can mean nothing to the reader," advising the author on the suitability of reconsidering the paragraph, and to confirm or correct it, something considerably more difficult but not substantially different from what the syntactic corrector all computers contain does. If computers had semantic analyzers, it could be good support for writers.

Anyway, good algorithms counting with the possibility of automatically designing what should be represented with the help of all the armamentarium of, at least, the classical and the fuzzy calculus, are currently unknown; in praxis, such meaning's analyzer would be a mechanical designer being able, in particular, to discriminate the intervening functions J. This is currently something really difficult to construct; if existing, it would maintain with the writer a short but "intelligent" conversation for which semantic understanding is necessary.

12.5. What can be done of a qualitative character without specifying numerical functions representing connectives and conditionals? To such a goal, which of the following statements,

p = "It is false that John is not very tall" and q = "It is false than John is not very short", is more true? The answer obviously depends on some data.

Let it be, for instance, $H(\text{John}) = H(J) \in (0, 2]$, the height in meters of John to which tall and short refer. Design membership functions μ_t, and μ_s, as those whose linguistic labels are, respectively, "tall" and "short" in (0, 2], and let's reasonably suppose μ_t is nondecreasing. Because short is an antonym of tall, define $\mu_s(x) = {}_t(2 - x)$. Modify these functions once affected by the linguistic modifier "very" by a function v: $[0, 1] \to [0, 1]$ that, like squaring them, is nondecreasing, and take a decreasing function f, such as $1 - Id$, to compute the degree of falseness, and a strong negation N for negation.

With all this pure symbolism, it is:

- $t(p) = f(N(v(\mu_t(H(J)))))$,

and

- $t(q) = f(N(v(\mu_s(H(J))))) = f(N(v(\mu_t(2 - H(J)))))$.

To compare these degrees it suffices to compare $H(J)$ and $2 - H(J)$:

(a) $H(J) \leq 2 - H(J) \Leftrightarrow H(J) \leq 1$, implies $\mu_t(H(J)) \leq \mu_t(2 - H(J)) \Rightarrow t(p) \leq t(q)$,

(b) $H(J) > 2 - H(J) \Leftrightarrow 1 < H(J)$, implies $\mu_s(H(J)) \geq \mu_s(2 - H(J)) \Rightarrow t(q) \leq t(p)$,

hence the conclusion is

$$\text{When } H(J) \leq 1, q \text{ is truer than } p, \text{and when } 1 < H(J), p \text{ is truer than } q.$$

Note that to arrive at this qualitative conclusion, only reasonable hypotheses on the involved membership functions and connectives should be supposed, and no particular specification of them is necessary. Nevertheless, for effectively computing the degrees $t(p)$ and $t(q)$, all these functions should be specified; for instance, with $f(x) = N(x) = 1 - x$, $v(x) = x^2$ and $\mu_t(x) = x/2$ with which it is $\mu_s(x) =$

$\mu_t(2 - x) = 1 - x/2$, it follows that $t(p) = (H(J)/2)^2$, and $t(q) = ((1 - H(J))/2)^2$, that if, for instance, were $H(J) = 1.8 > 1$ would give $t(p) = 0.36$, and $t(q) = 0.16$.

This example shows that with a not yet functionally specified symbolic representation, some qualitative conclusions can be reached, and that for computing actual numerical values, selecting and designing specific functional representations is necessary.

12.6. Because the pointwise ordering among membership functions in $[0, 1]^X$, $\mu \leq \sigma \Leftrightarrow \mu(x) \leq \sigma(x)$ for all x in X is transitive, the NC and EM "principles", respectively expressed in the forms $\mu \cdot \mu' \leq (\mu \cdot \mu')'$ and $(\mu + \mu')' \leq ((\mu + \mu')' = \mu + \mu'$, whenever the negation is strong, hold and have no possibility of failing. Consequently, it cannot be stated that fuzzy sets violate these principles.

Nevertheless, what in general is unknown is when they are, respectively, equivalent with the forms $\mu \cdot \mu' = \mu_0$ and $\mu + \mu' = \mu_1$, as happens in ortholattices; for such study, the functional expressible case offers some possibilities.

For instance, for NC and with continuous t-norms, t-conorms, and strong negations, the equivalence between the functional inequalities should be analyzed

$$T(a, N(a)) \leq N(T(a, N(a))), \text{and the equation } T(a, N(a)) = 0,$$

for all a in [0, 1], with which the inequality is obviously satisfied; but what about the reciprocal? The equation's solutions are known to be $T = W_f$ and $N \leq N_f$, satisfying the inequality; hence, these functions satisfy the equivalence.

Analogously, for EM the equivalence between

$$N(S(a, N(a))) \leq S(a, N(a)), \text{and } S(a, N(a)) = 1,$$

should be analyzed. Because it is clear that the second implies the first, and it is known that the solutions of the equation are $S = W^*_f$, and $N_f \leq N$, it is clear that these solutions are those for which the equivalence holds. Hence, it is with the triplets given by W_f, W^*_g, and N such that $N_g \leq N \leq N_f$ that both equivalences hold. In particular, they hold with $f = g$, in which case the possible negations are reduced to N_f.

In conclusion, in the case of imprecise words represented by membership functions of fuzzy sets, and due to the pointwise order transitivity, the new form of expressing the Aristotelian principles NC and EM by self-contradiction hold without exception. But this is not the case for the principles expressed in the old conjunctive/disjunctive forms that are typical of the classical logical calculus. Only with continuous t-norms, t-conorms, and strong negations is the problem's solution characterized and just proven equivalent to the first, but it remains open when the connectives are not functionally expressible, or they are but by means of functions in families other than continuous t-norms and t-conorms endowed with fewer laws than they have and strong negations.

References

1. C. Alsina, M.J. Frank, B. Schweizer, *Associative Functions. Triangular Norms and Copulas* (World Scientific, Singapore, 2006)
2. P. Klement, R. Mesiar, E. Pap, *Triangular Norms* (Springer, Berlin, 2000)
3. E. Trillas, L. Eciolaza, *Fuzzy Logic* (Springer, Berlin, 2015)
4. C. Alsina, E. Trillas, Ll. Valverde, On some logical connectives for fuzzy sets theory; J. Math. Anal Appl. **93**, 15–26 (1983)
5. H. Nguyen, E. Walker, *A First Course in Fuzzy Logic* (Chapman & Hall, Boca Raton, 2000)
6. H.-J. Zimmermann, *Fuzzy Set Theory and Its Applications* (Springer, New York, 2001)
7. E. Trillas, C. Alsina, J.M. Terricabras, Introducción a la lógica borrosa. Ariel, Barcelona
8. D. Dubois, H. Prade (eds.), *Fundamentals of Fuzzy Sets* (Kluwer, Dordrecht, 2000)
9. H.-J. Zimmermann, P. Zysno, Latent connectives in human decision making. Fuzzy Sets Syst. **4**, 37–51 (1980)
10. E. Trillas, S. Guadarrama, Fuzzy representations need a careful design. Int. J. Gener. Syst. **39**(3), 329–346 (2010)
11. B. Schweizer, A. Sklar, *Probabilistic Metric Spaces* (North Holland, New York, 1983)

Part II
Gathering Questions

(*) If the first part illustrated a sowing of ideas, this second part, perhaps to some readers' possible new discouragement, is just a harvest of questions neither posed, nor answered, in a complete form.

The author apologizes for his inability to present anything other than a dark jungle of more or less ambiguous ideas. Would some reader be interested in dissolving such darkness!

Chapter 13
A Few Questions on the Reasoning on Quantum Physics

The subject of this chapter concerns an overview of the model for representing the reasoning physicists conduct in their scrutiny of the microphysical world, that is, in quantum physics. Mainly, it is done for remarking that the laws of a mathematical representation's framework should be adapted to the requirements of the corresponding reality.

In that reasoning, the events are supposed to be precise, but uncertain, and involving facts observed thanks to sophisticated experiments, and without counting, in most cases, with all the information necessary for representing them well in a Boolean algebra. Of many quantum phenomena only their probability can be known.

Uncertainty is pervasive in most situations of quantum physics, and incompatibility is not always coincidental with contradiction; contradiction implies incompatibility, but not reciprocally. Also the distributive laws are not always valid, and the "quantum events" are not observable as they are, for instance, those corresponding to the movement of a satellite turning around the Earth; there are modifications the observation produces into what is observed. Its natural framework cannot be a Boolean algebra, even if some Boolean laws should be preserved.

13.1. The mathematical framework on which physicists reason, study, and represent the quantum phenomena, is an infinite-dimensional Hilbert vector space of functions *H*, with a scalar product, and whose vector subspaces correspond to the solutions of the significant equations and inequalities of quantum physics that are established thanks to some chosen operators. Hence the actual working universe is not *H*, but *S*(*H*), that of the vector subspaces of *H*; the statements of quantum physics are specified by the subspaces in *S*(*H*).

In *S*(*H*), the intersection of vector subspaces of *H*, that is also a vector subspace of *H*, contains the joint solutions of the equations and inequalities generating such subspaces, but it neither happens with their union, nor with the complement, that are not vector subspaces of *H*. Consequently, instead of the union and the

E. Trillas, *On the Logos: A Naïve View on Ordinary Reasoning and Fuzzy Logic*,
Studies in Fuzziness and Soft Computing 354, DOI 10.1007/978-3-319-56053-3_13

complement, the minimum vector subspace containing both subspaces, and the corresponding orthogonal subspace of the given one, respectively, should be taken.

That is, if A and B are two vector subspaces of H, their intersection $A \cap B$, their direct sum $A \oplus B$, and the orthogonal complement, $A^{\perp}$, $B^{\perp}$, and so on, should be taken for each of them for representing, respectively, the conjunction, disjunction, and negation of statements. The direct vector sum $A \oplus B$ contains the sums of all pairs of vectors in A and B, and the orthogonal complement $A^{\perp}$ contains all vectors $f^{\perp}$ orthogonal to those $f \in A$.

With that, the set $S(H)$ of all the vector subspaces of H shows the structure of an orthomodular lattice, that is, a nondistributive ortholattice in which for each pair A, $B \in S(H)$, there exists a "relative complement of A with respect to B", namely

$$B{-}A := B \cap A^{\perp}, \text{with which it is } A \subseteq B \Leftrightarrow B = A \oplus (B{-}A)$$

that, provided A and B were among the Boolean subspaces in $S(H)$ (reduced to that only containing the vector zero and H containing all vectors), would just coincide with $B = A \cup B$. Obviously $\{0\}$ is the orthogonal complement of H, and reciprocally.

The ordering of the algebraic structure generated by these operations is, obviously, the inclusion of vector subspaces, and because $S(H)$ is not a distributive lattice, it is not a Boolean algebra, and the law of perfect repartition $A = (A \cap B) \oplus (A \cap B^{\perp})$ does not universally hold.

All this leads to taking an abstract orthomodular lattice as a formal model for quantum reasoning; this is in the same vein as passing from sets to abstract Boolean algebras for the precise reasoning in the macroworld.

13.2. Notwithstanding, if the reasoning on quantum physics seems to be pretty well represented by the former model, introduced in 1936 by Garrett Birkhoff and John von Neumann, not all scholars today agree on it, and some of them advocate for a simpler one keeping fewer laws than orthomodular lattices; for instance, it is sustained by some authors that the law of excluded-middle, $p + p' = 1$, valid in orthomodular lattices, does not actually hold within quantum phenomena.

The model is not fully undisputed and, for instance, there is no general agreement on representing conditional statements, sometimes done by the so-called Sasaki arrow (or hook), $p \rightarrow_S q = p' + p \cdot q$. but also sometimes by the Dishkant arrow, $p \rightarrow_D q = q + p' \cdot q'$.

Each of these arrows is contrasymmetrical with the other:

$$q' \rightarrow_D p' = p' + (p')' \cdot (q')' = p' + p \cdot q = p \rightarrow_S q, \text{and, hence,} \; q' \rightarrow_S p' = p \rightarrow_D q;$$

both verify the modus ponens (MP) inequality, and, as said formerly, reduce to $p' + q$ with the Boolean elements of the orthomodular lattice. Contrary to the Boolean case, a maximum representation of the conditionals $p \rightarrow q$ satisfying the MP inequality does not exist; the Sasaki and the Dishkant arrows are only maximal elements among such representations, but verify: $p \leq q \Leftrightarrow p \rightarrow_S q = p \rightarrow_D q = 1$; that is, both "order" the lattice.

It should be pointed out how, by simply rejecting the distributive laws, the resulting calculus is notably different from the Boolean one; it corresponds, indeed, to a different reality.

13.3. In the case of orthomodular lattices, contradiction implies incompatibility, but not always reciprocally. In fact, $p \leq q' \Rightarrow p \cdot q = 0$, but for proving the reciprocal the verification of $p = p \cdot q + p \cdot q'$, the law of perfect repartition between p and q from which $p \cdot q = 0$ implies $p = p \cdot q' \Leftrightarrow p \leq q'$ is sufficient. Thus, and different from the classical case, there can exist pairs of objects that, being incompatible, are not contradictory; incompatibility affects different pairs than contradiction does, and contradiction seems to appear only once incompatibility is constrained by perfect repartition, something that corresponds with some observational facts in quantum physics.

Whatever mathematical model is chosen for representing quantum reasoning, it should allow capturing the mentioned differences with classical reasoning.

Anyway, there is still another subject not currently well captured that corresponds to the uncertainty pervading quantum phenomena and that is revealed by experimentation. This is a subject that, in the classical case, is solved once presuming that partitions are always possible by incompatibility of events, and by adding the additive law of probabilities under the hypotheses that they are measures of uncertainty, by identifying, at the end, and in the corresponding language, "p is uncertain" with "p is probable", by reducing their qualitative meanings to the partial lattice's order. But there are now incompatible events not being contradictory, something making things more complicated and leading to a restricted definition of the probability's additive law, by:

$$\text{Provided } p \leq q', \text{then } \text{prob}(p+q) = \text{prob}(p) + \text{prob}(q),$$

that, jointly with the axiom prob(1) = 1 allows us to prove prob(0) = 0, because 0 is contradictory with 1, $0' = 1$, and prob(0) = prob(0 + 0) = prob(0) + prob(0), and also to prove $\text{prob}(p') = 1 - \text{prob}(p)$, because from the contradiction of p and p', $p = (p')'$, it follows that $\text{prob}(p + p') = 1 = \text{prob}(p) + \text{prob}(p')$. With this slight change of the additive law, the mapping prob preserves the basic laws of probability.

Although $p \cdot q \leq q \leq q + p' = (p \cdot q')'$, that is, $p \cdot q$ and $p \cdot q'$ are contradictory, it cannot be implied that $\text{prob}(p) = \text{prob}(p \cdot q) + \text{prob}(p \cdot q')$, because it is not sure that $p \cdot q + p \cdot q'$ coincides with p; hence knowing the probabilities of $p \cdot q$ and $p \cdot q'$ does not allow us to know the probability of p. This makes analyzing the uncertainty of p from the uncertainty of its interaction with q difficult, and shows a basic difficulty for proving that conditional probability is indeed a probability in the universe restricted to the traces of q, as done in Kolmogorov theory by defining $\text{prob}(p/q) = \text{prob}(p \cdot q)/\text{prob}(q)$ provided $\text{prob}(q) > 0$.

Summing up, if "quantum" probability can be defined on an orthomodular lattice by a slight modification of its additive law, conditional probability requires a different view, and it is at such a point that the theoretic modeling of quantum reasoning still remains a subject without general agreement.

Notwithstanding, all that shows that for analyzing the reasoning of quantum physics weaker structures than Boolean algebras are necessary, and it reinforces the idea that each specific type of reasoning deserves to be represented in a suitable mathematical framework in which the use of formal laws without a real counterpart should be excluded.

No liable conclusions can be reached by presuming, for instance, the distributive laws, once it is known that they cannot hold in a reasoning whose statements are represented by the vector subspaces of a Hilbert space, and, in consequence, don't allow knowing what is in a subspace A by knowing what it shares with any B and its orthogonal subspace. Of course, the same will happen with a reasoning whose statements are represented by the vector subspaces of the n-dimensional vector space R^n, inasmuch as the distributive laws also fail in it.

Before adopting a mathematical model for representing some reasoning in a given universe, each law only can be accepted after testing it against the reality of the considered statements on such a universe; it is analogous to accept the commutative law of conjunction with statements whose conjunctions cannot actually commute or, with some difference, to consider the exclusive disjunction $p\Delta q = (p + q) \cdot (p \cdot q)'$ instead of the inclusive one $p + q$, even if both commute.

References

1. G. Bodiou, *Théorie Diallectique des Probabilités* (Gauthier-Villars, Paris, 1962)
2. L. Beran, *Orthomodular Lattices* (Reidel, Dordrecht, 1985)
3. G. Birkhoff, J. von Neumann, The logic of quantum mechanics. Ann. Math. **37**(4), 823–843 (1936)
4. G. Cattaneo, L. Laudisa, Axiomatic unsharp quantum theory (from Mackey to Ludwig and Piron). Found. Phys. **24**, 631–683 (1994)
5. G. Ludwig, *An Axiomatic Basis for Quantum Mechanics* (Springer, Berlin, 1985)
6. G. Birkhoff, *Lattice Theory* (Pubs, AMS, New York, 1948)

Chapter 14
Questions on Uncertain, Possible, and Probable

In plain language, the predicative words "possible" and "probable" are often used synonymously, but without argument for it, and what seems more suitable is to say of something probable that it is possible, but not reciprocally. In language, "possible" seems to be more largely applicable than "probable". For instance, if when throwing a single die it is probable to get five points, it is clear that it can be expected because this output is among "those that are possible"; obtaining eleven points is not possible, and consequently there is no sense in attributing to it the property of being probable. In addition, there are also life's ordinary situations linguistically qualified as "possible but not probable", even if they can actually have a very small probability. For instance, it is possible that in a few minutes my old friend John, 10 years older than I am, and from whom I have heard nothing in the last 10 years, can call me by phone, but it deserves to be qualified as something improbable or, at most and if John is still alive, with a very small probability provided it could be effectively computed.

The theoretical distinction between possible and probable can be seen, in principle, in the different axioms with which the mathematical theories on the measures of probability by Kolmogorov in 1933, and on the measures of possibility by Zadeh in 1978, formalized the measuring of probability and possibility, respectively. Notwithstanding, these theories suppose that the elements to which the predicative words probable and possible are applied, belong to actually strong types of lattices with a negation, something that, in plain language and as was said, is odd, risky, and even dangerous to always suppose it. Let's begin with an overview of these theories that, nevertheless, refer to the concept of probability and possibility but not, directly, to the use of the words "probable" and "possible" in plain language but, at most, in some particular and specialized part of it.

E. Trillas, *On the Logos: A Naïve View on Ordinary Reasoning and Fuzzy Logic*,
Studies in Fuzziness and Soft Computing 354, DOI 10.1007/978-3-319-56053-3_14

14.1. Kolmogorov established his theory of probability on the following hypotheses:

1. A measure of probability, prob, assigns a number between 0 and 1, to "events" represented by subsets in a Boolean algebra on a set Ω included in the power-set 2^X of some universe X, prob: $\Omega \to [0, 1]$, and such that
2. prob(Ω) = 1.

This presumes that all the laws valid in the Boolean algebra Ω are applicable to the "events", something supposed as actually happening; in particular, it is supposed the existence of perfect classifications, or partitions, of the nonempty subsets in Ω containing several elements.

3. The essential axiom for the mapping prob is its additive law,

$$\text{prob}(A \cup B) = \text{prob}(A) + \text{prob}(B),$$

provided the intersection of the "events" A and B were empty, $A \cap B = \emptyset$; that is, $\{A, B\}$ is a perfect classification or partition of $A \cup B$.

From this follows prob(A') = 1 − prob(A), and prob($\emptyset$) = 0. With it, is proven that prob is a measure for the graph (Ω, $\subseteq$) with the minimum $\emptyset$, and the maximum Ω, and thanks to the existence of relative complements in the Boolean algebras. That is, and as formerly proven, $A \subseteq B$ => prob(A) $\leq$ prob(B).

Thus, probabilities can be seen in Boolean algebras as (additive) measures of the predicative word "probable" applied to the events, and by supposing that $\emptyset$ is the less probable subset, Ω the most probable of them, and that "A is less probable than B" is just done by $A \subseteq B$, identifying $<_{\text{probable}}$ with $\subseteq$; that (with finite subsets) "less elements" is equivalent to "less probable", with $\emptyset$ minimal, and Ω maximal.

All that comes from the fact that, genetically, probabilities come from problems in which the "represented events" can not only be precisely named and counted, but also perfectly classified into crisp parts. In addition, in Kolmogorov's interpretation of probability, it should be taken into account that events are not only considered random ones, in the sense of being obtainable by indefinitely repeating an experiment under exactly the same conditions at each repetition, but that frequencies of its appearances can be somehow computed.

Kolmogorov's theory arises from a crisp objectivistic interpretation of probability through random experiments; in itself, it is but an abstract mathematical theory of (normalized) additive measures, and, today is not only the most widely known interpretation of probability, but is also responsible for most of the applicative successes of mathematical statistics. The interpretation by Kolmogorov reflects the famous and wise statement, "Nothing is more practical than a good theory."

Notwithstanding, not all the successful measures in the applications are additive; for instance, the big family of Sugeno's λ-measures, m, verifying:

$$m(A \cup B) = m(A) + m(B) + \lambda.m(A).m(B), \quad \text{with} \; -1 < \lambda, \quad \text{if } A \cap B = \emptyset,$$

contains additive measures if it is $\lambda = 0$, but with $\lambda < 0$ has super additive and with $\lambda > 0$ has subadditive measures.

In "reality" there are relevant situations in which information, or evidence, is not so precise for allowing its splitting in separate pieces, and it should be remarked that, in language, the term "probable" is not only applied to precise, but also to imprecise, statements that cannot be represented by sets, a case in which the immediate applicability of the Kolmogorov theory is at least dubious; the existence of crisp partitions is so in some nebulous situations where what appears is not a clearly separable mixing of information. Sometimes the measure of a "totality" is greater than the sum of the measures of its parts, and sometimes it is less. There are also "events" whose repetition under the same exact conditions is not possible; even in some situations, former instances mean nothing for the next one.

14.2. Zadeh established his theory on the measures of possibility with fuzzy sets in the unique basic fuzzy algebra (BAF) that is a lattice, the De Morgan algebra, ($[0, 1]^X$; min, max, $1 - \text{id}$); but, namely and for what concerns its basic hypotheses, it can be considered in an abstract De Morgan algebra (M; $\leq$; 0, 1; $\cdot$, +;$'$), where the order $\leq$ is that of the lattice, and supposing that all its elements can be qualified as "possible". These hypotheses are just the following.

A measure of possibility is a mapping π: $M \to [0, 1]$, such that:

- $\pi(0) = 0$,
- $\pi(1) = 1$, and
- $\pi(a + b) = \max(\pi(a), \pi(b))$, for any pair a, b in M, regardless of being $a \cdot b = 0$, or not.

Because $a \leq b$ is equivalent to $a + b = b$, it follows that $\pi(b) = \pi(a + b) = \max(\pi(a), \pi(b)) \geq \pi(a)$; thus, π can be seen as a measure of the predicative word "possible" when applied to the elements in a De Morgan algebra, and with qualitative meaning (M, $\leq$), once the relation $\leq_{\text{possible}}$ is identified with the lattice's order $\leq$ of the De Morgan algebra.

Note that π is not additive but subadditive, because it verifies $\pi(a + b) = \max(\pi(a), \pi(b)) \leq \pi(a) + \pi(b)$; totalities measure less than their parts. Regarding $\pi(a')$, it not always holds its equality with $1 - \pi(a)$, that cannot be a law, because for any Boolean element a in M (i.e., verifying $a + a' = 1$), it follows that $\pi(a + a') = \pi(1) = 1 = \max(\pi(a), \pi(a')) \leq \pi(a) + \pi(a')$, and thus, $1 - \pi(a) \leq \pi(a')$.

An advantage of Zadeh's theory is that it models the applicability of "possible" to elements that do not need to be in a Boolean algebra, as is the case with membership functions of fuzzy sets, and for which in principle no crisp partition exists; neither the existence of crisp partitions, nor that the negation is a Boolean complement, is previously supposed. A serious disadvantage is, nevertheless and for plain language, that De Morgan algebras are still too strong algebraic structures with a lattice basis, in which some of their properties, such as they are the conjunction's commutative law and the distributive laws, cannot always be presumed in language.

In both Boolean and De Morgan algebras, the weight of laws of a syntactic origin is actually too strong for language; this occurs, for example, in Boolean algebras with the law of "perfect repartition", $a = a \cdot b + a \cdot b'$. It is a law that neither holds in all ortholattices, nor in De Morgan algebras and that, as shown in the BAFs, forces that conjunction and disjunction are not dually linked. In addition, double negation, $(a')' = a$, and duality $(a + b)' = a' \cdot b'$, are laws not always verifiable in plain language. In De Morgan algebras, distributive laws are among those that cannot always be supposed in plain language.

Hence, both theories of Kolmogorov and Zadeh are only applicable to those parts of language in which their presumed laws can be accepted. Both presume that the relations $<_{\text{probable}}$ and $<_{\text{possible}}$ coincide with the partial order of the corresponding lattice.

Anyway, Zadeh's theory is not fully objectivistic as is Kolmogorov's theory; it neither supposes nor excludes that the statements whose possibility is to be measured should represent events obtainable in experiments repeatable under the same conditions; it admits events for which just some precise or imprecise information is known and can be represented by statements generating linguistic collectives or fuzzy sets. It does not force that statements should be endowed with an ortholattice algebraic structure, even if supposing it is a De Morgan one where some strong laws hold, such as the distributive ones. It proceeds through a different weakening of Boolean algebras than the quantum calculus does.

In addition, and as is well known, a membership function taking the value 1 at some point can be interpreted as a distribution of possibility conditioned by the previously available information on the use of its linguistic label but not, in general, as a probability distribution.

Were each fuzzy set's membership function μ equal to a probability p_μ (in a universe X being a Boolean algebra, or in an orthomodular lattice), because

$\mu \leq \lambda \Leftrightarrow \mu(x) \leq \lambda(x)$, for all x in X, it is also $\mu(x') \leq \lambda(x') \Leftrightarrow 1 - p_\mu(x) \leq 1 - p_\lambda(x) \Leftrightarrow p_\lambda(x) \leq p_\mu(x)$, or $\lambda(x) \leq \mu(x) \Leftrightarrow \lambda \leq \mu$,

it would follow that $\mu = \lambda$; that is, the pointwise ordering between these fuzzy sets collapses in the identity. It follows a rare ordering of membership functions, under which two of them only can be coincidental or not comparable. With it, many useful applications of fuzzy sets would be lost. The pointwise ordering is not a natural form for ordering probability measures. Interpreting fuzzy sets in a Boolean algebra, or in an orthomodular lattice, as a measure of probability just leads to a very odd "theory" for both fuzzy sets and probabilities.

Notwithstanding, this does not mean that each particular numerical value of a fuzzy set cannot be obtained as the value of a probability. For some concretion, given the finite fuzzy set in $X = \{1, 2, 3, 4\}$,

$$\mu = 0.5/1 + 0.7/2 + 1/3 + 0/4,$$

there are many quadruplets (p_1, p_2, p_3, p_4) of probabilities, each able to give the corresponding value of μ; for instance:

$$\mu(1) = 0.5 = p_1(1); p_2(2) = 0.3; p_1(3) = 0.2; p_1(4) = 0,$$
$$p_2(1) = 0.2; \mu(2) = p_2(2) = 0.7; p_2(3) = 0; p_2(4) = 0.1,$$
$$p_3(1) = 0; p_3(2) = 0; \mu(3) = p_3(3) = 1; p_3(4) = 0,$$
$$p_4(1) = 0.5; p_4(2) = 0.2; p_4(3) = 0.3; \mu(4) = p_4(4) = 0.$$

This simple example shows that there is nothing against those cases in which each numerical value of the membership function comes from a specific random variable, that for some fuzzy sets it seems possible to design a series of random experiments from whose respective probabilities the values of its membership function can be obtained; something that, in general, is impossible by means of a single probability and corresponds to a statistical view based on managing random variables. Anyway, what is not clear enough is on which characteristics of the fuzzy set's context, such random experiments, or random variables, could be linked; it seems dependent on how the contextual information could be acquired.

In short, what cannot be excluded at all is the possibility of obtaining the values of a fuzzy set's membership function through a statistical methodology. But, in any case, there cannot be coincidence between a fuzzy set and a probability defined on the same ground. The theories of fuzzy sets and probabilities have different goals, although fuzzy sets and possibility measures have closer goals.

14.3. As has been shown, Kolmogorov's theory can be imported (with some modifications) into algebraic structures weaker than Boolean algebras, as it is into orthomodular lattices in the so-called "quantum probability calculus," corresponding to a particular form of language in which statements are precise even if not all the information on them is known, and where, instead of crisp partitions $p = q + r$ with $q \cdot r = 0$, it is supposed with $q \leq r'$ that, only provided q and r were Boolean elements, would coincide with $q \cdot r = 0$; in orthomodular lattices, contradiction ($q \leq r'$) implies incompatibility ($q \cdot r = 0$), but not reciprocally. These are two concepts only coincidental in Boolean algebras and thanks to its basic perfect repartition law, but that in language (and even in most algebras of fuzzy sets) are actually independent of each other. It should be pointed out that the perfect repartition law comes directly from distributivity:

$$1 = p + p' = > q = q \cdot (p + p') = q \cdot p + q \cdot p'$$

and that distributivity is not valid in orthomodular lattices; perfect repartition is also not always valid in De Morgan algebras because in them $p + p' = 1$ is not a law.

The quantum case shows how the information available on what is stated affects the laws that can be supposed, but it should be noted that, in plain language, they cannot be supposed to hold and that, when dealing with one of its parts, some previous checking of them is necessary. It can be said that fuzzy logic is the first approach breaking, in language, the usually supposed lattice structure of statements; something reflected, for instance, in the many applications where the conjunction is represented by the product instead of the minimum, and for which the usually presumed syntactic property $p \cdot p = p$ is lost.

In addition, it is also worth noting that in the so-called intuitionistic lattices (in which negation is not strong) a kind of relational, or conditional, probability can be defined, but that (as was proven) such lattices are included through a previous equivalence in a Boolean algebra of classes.

14.4. Once a short review of the mathematical theories of probability and possibility is done, let's overview the uses of possible, probable, and uncertain, in plain language.

As has been repeatedly said, in general, plain language is not submitted to all the laws of algebraic structures such as the algebras of Boole, De Morgan, and the like. Introducing an algebraic structure and formalizing it means constraining plain language to an artificial one. Plain language is almost all we count with to express what we experience with the senses, and what we elaborate intellectually; it has great and essential flexibility, and for "structuring language" it is necessary to consider only some of its specific parts, submit them to a regimentation, forget some particularities, and finally adopt a mathematical model that, like all models, at the end is but a more or less valid simplification of reality. In the particular case of a lattice's model, it is extremely rigid; for instance, it supposes the conjunction to be the greatest statement among all those from which the two statements submitted to conjunction follow forward. It also supposes some laws that are not always valid for statements such as the associative laws of conjunction and disjunction, or the existence of two statements considered as "neutral" and "absorbent" for the conjunction, and presuming that a clear and general concept of truth and falsehood is known in the corresponding piece of language. Syntax is important, but what really matters in language is semantics; a badly syntactically constructed linguistic statement is often comprehended well enough, but confusion of meaning is what actually produces a serious lack of comprehension.

What follows tries to shed some light on the words *uncertain*, *possible*, and *probable*, but without imposing too many laws and towards a not still existing total scientific domestication of the uses of these words in plain language, something for which there is yet a lack of practical knowledge.

14.5. What does it mean that something can be qualified as uncertain in plain language? Uncertain, as the opposite word of certain, implies (without a necessary equivalence) not-certain; hence, uncertain refers to some lack of certainty with respect to aspects surrounding what is stated but, instead, not-certain means a total lack of certainty. For instance, the statement, "It is uncertain that candidate C will win the election," just means that the statement "Candidate C will win the election" is, for what is contextually and currently known as certain, not necessary and unsafe; that, if betting on it, any of those knowing what surrounds the election will risk a very low wager by believing that the bet can be easily lost. It is not a properly objectivistic view, but a subjective one, based on the experience of the player.

Anyway, for capturing a full-meaning of the predicative word U = uncertain in a universe X of statements p, q, r and so on, a quantity $(X, <_{\mathrm{U}}, m_{\mathrm{U}})$ should be specified, and even an analysis of what certain means seems to be previously necessary for linking its qualitative meaning with the opposite word uncertain in the form $<_{\text{uncertain}} = <_{\mathrm{U}} = <_{\text{certain}}^{-1}$.

Note that the coincidence of the relation $<_{\mathrm{U}}$ (less uncertain than) with the order of a lattice (be it a Boolean, a De Morgan algebra, or whatever ordered structure) is but a supposition involving the hypothesis that X can be endowed with such an algebraic structure, something on which one must be, for what has been commented, extremely cautious and check, previously, if each of the laws defining such a structure can actually be supposed. To specify a quantity reflecting a full-meaning of U, and as happens with all words, some "experience" of the use of uncertain in plain language is necessary for establishing, at least, the empirical relation "less uncertain than", knowing which statements are maximal, which are minimal, and specifying a measure.

It can be supposed that in language the word uncertain inherits a "history relative to its use," coming from some experiences linked with contexts in which uncertain has been formerly used, and acquired either directly or by means of some "contagious" contact, oral or written, with others using it.

The relation $<_{\mathrm{U}}$ will always depend on the context surrounding the use of U and, for instance, supposing it coincides with the order of a lattice (always verifying $p \leq q \Leftrightarrow p + q = q \Leftrightarrow p \cdot q = p$) could be very risky because, for instance, in language and due to time intervention, it can be $p \cdot q \neq q \cdot p$. Additionally, in language the exclusive disjunction is often managed $p \,\Delta\, q = (p + q) \cdot (p \cdot q)'$, instead of the inclusive disjunction $p + q$, but $p + q = q$ implies $p \,\Delta q = q \cdot (p \cdot q)'$ that, in a Boolean algebra means $p \,\Delta\, q = q \cdot p'$, and that because $q \cdot p' \leq p'$ is $p \,\Delta\, q$ contradictory with p and only coincides with q if it is $q \leq p'$, that is, if p and q are contradictory.

Inasmuch as no specific and satisfactory general theory of uncertainty comprising its measuring is currently well known, there is room for interpreting the word *uncertain* in the form each can be contextually able to do. For instance, those mastering probability theory or possibility theory tend to identify uncertain with probable or, respectively, with possible, even if obviously uncertain is more general than probable and possible; something probable or possible is uncertain, but the reverses are not clear enough. For instance, under which random experiment can a probability be computed for "Candidate C will win the election"? Of course, there is neither a way of designing experiments such as that of throwing a die, nor of considering a Boolean or a De Morgan algebra containing the presumed results of such an experiment, and less again ways of repeating it exactly under the same conditions at each repetition; for instance, a former election even with the same candidates will not present the same surroundings of the next and, at least, some ideal similarities should be chosen.

It is debatable if Kolmogorov theory is directly applicable to this kind of "events", and, for instance, Zadeh introduced a calculus of probabilities for imprecise events represented by fuzzy sets but showing difficulties for defining conditional probability, such as also happens with quantum probability. In addition, and in the case of precise words, the identification of $<_U$ with the inclusion relation among subsets $\subseteq$ leads to identifying "less uncertain than" with "less elements than", perhaps a too risky identification. Possibility theory that involves subjective views could be more suitable even if its results can be subjected to a large lack of certainty.

14.6. Regarding the use of P = "probable" in language, for capturing its full-meaning the relation $<_P = <_{probable}$ should be captured specifying a measure m_P. What is obvious is that when the considered statements are imprecise, they cannot constitute a Boolean algebra, nor an orthomodular lattice, but that what is open is, in some cases, a De Morgan algebra, and even a different BAF. When the statements are precise, a Boolean algebra of sets representing them exists and hence what should be debated is if relation $<_P$ is, or is not, the inclusion of sets ($\subseteq$), and if there exists an additive measure m_P. What can be easily accepted is that $\subseteq$ is contained in $<_P$, but the reciprocal is in general dubious; thus, because $A \subseteq B$ implies $A <_P B$, it follows that $m_P(A) \leq m_P(B)$, provided m were a measure for $<_P$; $<_P$—measures are $\subseteq$—measures, but it could not be guaranteed that a $\subseteq$—measure is always $a <_P$—measure. Hence, it is not for sure than a probability could always, and in plain language, be a good enough measure of the word "probable", and less again that such measures should be additive.

For instance, to identify what "It is probable that candidate C will win the election" (shortened to "C is P") means it is first necessary to know the relationships "C is P is less probable than D is P", for all the candidates D in the election process; and once the graph (Candidates, $<_P$) and its maximal and minimal elements are known, more information is required on the context surrounding the election for specifying a measure. This does not mean, of course, that such a measure cannot be estimated by a statistical methodology.

There is a basic difference between $<_P$ and the relation of "comparative probability", introduced by *T.* Fine; such a difference lies in that comparative probability is a total or linear order, but $<_P$ cannot be always supposed to be so. Perhaps Fine's relation could be viewed as an extension of $<_P$; it deserves further study, but seems related to the linear relation of working meaning $<_m$ previously introduced for measures, and enlarging $<_P$.

14.7. Of course, similar comments can be made concerning the word $\prod$ = possible, as modeled in Zadeh's theory of possibility, and the use of the word $\prod$ in language by a full-meaning specified by a quantity $(X, <_\prod, m_\prod)$ in which neither $<_\prod$ should necessarily coincide with the order of a De Morgan algebra, nor $m_\prod$

necessarily verifies the axioms of a possibility measure. Namely, that it does not always grow under the max operation.

Even accepting that "if probable, then possible, but not reciprocally", it does not imply that given a measure of probability prob and one of possibility poss, it should always be prob(x) $\leq$ poss(x), for all x in a universe where both measures prob and poss are defined; it can be easily checked in a finite universe. For this reason, Zadeh introduced the concept of consistent pairs (poss, prob): such a pair of measures is consistent if it is prob $\leq$ poss.

Such a concept is in agreement with the chain of inclusions, $<_P \subseteq <_\prod \subseteq <_U$, concerning the qualitative meanings of the three words, and also with the typical expression "if it is probable, then it is possible", and in any case, "it is uncertain". In a finite Boolean algebra it is easy to prove there are probabilities that coincide with possibilities, but it is limited to degenerate measures, that is, those only taking the values 0 and 1; essentially, probabilities and possibilities are different measures. To specify a probability more information is needed than for specifying a possibility. $m_P \leq m_\prod \leq m_U$ seems that it could be a coherence relation among the respective measures.

14.8. The ordinary use in language of the words *uncertain*, *possible*, and *probable*, is not yet well known as is that of the word *big* in a closed interval of the real line; it requires a kind of experimental work that, for instance, could help to shed some light into the debate on the different views of probability between the Kolmogorovians, frequentialists, or objectivists, and the Bayesians or subjectivists. If the first base their view in the previous application of probability to precise words denoting random events represented in Boolean algebras of crisp sets, the second even try to apply probability to imprecise words neither denoting random events nor representable in those algebras, nor in more general ortholattices; Bayesians seem to be closer to the use of "probable" in plain language and ordinary reasoning than are Kolmogorovians.

The essential point for objectivistic interpretation lies in decomposing events into disjoint events to allow accepting the additive law of probability. But, with imprecise events represented by membership functions of fuzzy sets, the situation is different because having a decomposition $\mu = \alpha + \beta$ with $\alpha \cdot \beta = \mu_0$ (the function constantly equal to zero, representing the empty set) implies working in a BAF that, if functionally expressible, requires solving the functional system of equations $a = G(b, c)$ and $F(b, c) = 0$.

For instance, in a standard algebra $([0, 1]^X, F, G, N)$, where G is a continuous t-conorm, F a continuous t-norm, and N a strong negation function, F should be a t-norm in the family of Lukasiewicz, $F = W_\Delta$ (with Δ: $[0, 1] \to [0, 1]$ an order automorphism), because these continuous t-norms are the only ones with zero-divisors, and then $\Delta^{-1}(\max(0, \Delta(x) + \Delta(y) - 1)) = 0$ means $\Delta(y) \leq 1 - \Delta(x)$, or $y \leq N_\Delta(x)$, implying that the fuzzy sets α and β should be contradictory ($\beta \leq \alpha'$) in respect to the negation N_Δ.

That is, for counting with such kinds of "partitions", a very particular algebraic structure seems to be necessary; note that t-norms W_Δ are the only t-norms verifying the principle of noncontradiction by conjunction ($\alpha \cdot \alpha' = \mu_0$) among the standard BAFs. These algebras are not lattices except if $F = \min$ and $G = \max$, in which case they are De Morgan algebras. With these structures, one of the problems in plain language is the identification, inside a current problem, of a statement playing the role of "absorbent" for the conjunction, that is, being represented by the fuzzy set with membership function μ_0.

All the above needs to be searched in language through designing controlled processes of experimentation, and checking what is observed against some of the known mathematical models, as well as for getting hints towards establishing new models. For it, a search of the Web, as done in a recent paper by Sergio Guadarrama, Eloy Renedo, and myself, for studying how the linguistic conjunction "and" is actually used in language, could be a useful starting methodology. Without systematic observation in language, controlled experimentation, and adapting mathematical models to it, everything is but a kind of play with abstract ideas, perhaps of wishful thinking. It is difficult to imagine how a purely abstract formal kind of reasoning could arrive at, for instance, thermodynamics, and without not blind observation and controlled experiments. Our subject is empirical, and formalism is for helping its comprehension and for allowing computations.

Perfect partitioning is anchored in liking absolutes; a crisp set's partition suffices to name its parts by precise words specifying them. Perfect classification is an epitome for a "principle" of separation or isolation; that is, the belief that everything that is supposed to be composed can actually be separated in isolated parts whose union constitutes the composed totality, and that things grow by a clear superposition of pieces. It is an ideal that goes well with a lattice-type conception of language: a rigid conception that is artificial because, for instance, in language (and in life) time intervenes and situations are not always static; language is essentially dynamic. For instance, language also contains the opposites of words that, being several, sometimes are not comparable among them, and not coincidental with their negation; sometimes there is interplay with the given word, and not allowing perfect classifications. Zadeh's linguistic variables are a good example of it.

In language and reality there are many cloudy, gaseous like appearances, for which a principle of crisp separation is difficult to conceive, unless some crisp fixation, or representation of them, is accepted and that, in fact, does not appear in such a reality. Reality should be seen as it is, and represented in forms as close as possible to what it actually is; confusing name and thing is, jointly with believing that a name defines a real thing, just a dangerous "philosophical nominalism" based on believing that the universe of discourse can be perfectly classified in those x such that "x is P", and those such that "x is not P", whose intersection is empty, in which each set is isolated from the others. Paraphrasing Luigi Pirandello, a scientist should escape from words in search of where they can be applied; this could be metaphysics, but it is not science.

As commented before, what cannot be excluded is to modify measures by assigning, to what is to be measured, not a number in the real line but a mathematical object in a not linearly structured set. One reason for such a possible modification comes from the mentioned fact that often $<_P$ is not linear, but $<_m$ is always so. For instance, and on one hand, it could be thought of as assigning to the measures values being complex numbers, or intervals in the real line, with which $<_m$ ceases to be linear and more possibilities for its coincidence with $<_P$ are open, although no safety on it can be stated. However, it often also happens that the available additional information on the behavior of P only allows us to recognize at each x in X an upper limit (b_x) and a lower limit (a_x) of the values m can take at point x, and then, defining $m(x) = [a_x, b_x]$, or $m(x) = a_x + ib_x$, cannot seem bizarre.

This comment opens the door to consider (when it can be suitable) what, in fuzzy set theory, are called type-two fuzzy sets, consisting in assigning fuzzy numbers in [0, 1] instead of crisp numbers in the unit interval; of course, crisp numbers and intervals are particular cases of fuzzy numbers and, in addition, type-two fuzzy sets are not only more general, but can immediately represent usual linguistic statements such as "the measure is high", by representing the linguistic label "high" as a fuzzy number in [0, 1], that is, by a membership function μ_{high} of the linguistic label "high" in such an interval, and as they are, for instance,

$$\mu_{\text{high}}(x) = x,$$

or

$$\mu_{\text{high}}(x) = 0 \quad \text{if} \quad 0 \leq x < 0.8,$$

and

$$\mu_{\text{high}}(x) = 1 \quad \text{if} \quad 0.8 \leq x < 1,$$

with the second equivalent to the interval [0.8, 1].

With this, several forms of interpreting and representing "high" are possible, and open a window not only for representing and measuring more statements in natural language, but also perhaps for enlarging the application of probabilities when both its arguments and values are themselves imprecise or uncertain. This is, for instance, the case of the typical linguistic statement, "It is with high probability that John is rich," that could be translated into fuzzy terms by

$$\text{prob}(\mu_{\text{rich}}(\text{John})) = \mu_{\text{high}},$$

once a theory of probability for imprecise events and whose values are fuzzy numbers can be established. Nevertheless, the laws under which such "prob" should be defined are not actually known. This is a very important topic towards computing with words.

14.9. This chapter is not, by its own nature and like the others in this book's Part II, a conclusive one. It only offers a reflection for exciting its potential readers' interest through some patterns relative to a new view of the meaning of words in plain language, and without conceiving this concept as one crisply definable by necessary and sufficient conditions, but by quantities each specifiable through the contextual available information on the action of the word in the universe of discourse and, also, by adding reasonable hypotheses on it when necessary for its design.

In this view, meaning is not seen as a universal concept, isolated from both the context in which the word is used and the purpose of such use. The meaning of a word in a universe of discourse is not unique, but there are many possible meanings in each universe and, consequently, what cannot be thought is a single and universal meaning of the words *uncertain*, *possible*, and *probable*, like there is not a single probability for the events appearing in throwing a die, but several depending on the information available and acquired on the die by, for instance, through a nondestructive analysis of it.

There is uncertainty even in specifying a measure, be it uncertain, or possible, or probable. Almost everything in plain language is endowed with uncertainty, and a point on which there is a coincidence between what is presented in this chapter and the Bayesian theory of probability is that everything depends, at least, on the previously available information on the subject currently under study. Measures depend on some a priori information on what surrounds the use of the corresponding word; indeed, each measure can be viewed as "conditioned" by such information, and, when it changes, an updating of what has been previously presumed is necessary. A relevant difference between the Bayesian approach to probability, and the meaning concept in plain language, as it is approached here, is that the first mainly refers to events admitted as something that could actually happen, but meaning refers to the use of words in language that, often enough, not only refer to "physical" situations, but to virtual or informative ones, and that can count with a history of social uses. If defining a probability requires a strong algebraic structure, in plain language it does not seem to exist in a universal way; in each case, and as fuzzy logic shows, a particular algebraic structure should be searched for.

What is not done here is the more than debatable identification of uncertainty and risk; the risk taken when acting with uncertain knowledge is not considered as it properly belongs to the field of decision theory.

In sum, language is extremely complex; in the era of information it is, perhaps, the most complex system computer science is faced with, and that for comprehending well what information is, needs to be scientifically domesticated inasmuch as information is basically conveyed by a language, be it plain or artificial, and employed for reasoning.

14.10. A comment trying to approach the quantity's model of meaning to the Bayesian interpretation of probability is still in order, even if it remains just as a naïve trial.

In such an interpretation, and once a priori nonnull probabilities $p(a)$ and $p(b)$ are established, the probability of a can be updated after knowing that of b, by means of the well-known Bayes' formula, $p(a/b) = p(a \cdot b)/p(b)$, in which the hidden conditional "If b, then a", is conjunctively represented by $a \cdot b$, and not by $b' + a$, as is always done in Boolean algebras but that does not allow $p(./b)$ to be a probability. It shows that the interpretation of the formula does not properly concern the formal Boolean language of the model, but to something external to the model, that is, in the corresponding plain language. Note that it is $p(a/b) \leq p(a) \Leftrightarrow p(a \cdot b) \leq p(a).\ p(b)$ and $p(a) < p(a/b) \Leftrightarrow p\ (a \cdot b) > p(a).\ p(b)$. In any case, it seems that Bayesians could keep some linking with the presented conception of meaning and hence it deserves to be explored a little bit further.

Suppose a word P in X, with meaning given by a quantity $(X, <_P, m_P)$. Suppose that there is a word Q, with meaning $(X, <_Q, m_Q)$, such that it can be asserted "If x is Q, then x is P", for all x in X. Once this is known, is the a priori meaning of P affected? Provided the answer were affirmative, how would it be affected? Is there a new quantity $(X, \overset{*}{<}_P, \overset{*}{m}_P)$, giving an a posteriori meaning, conditioned by the information the conditional can furnish? How is such a new meaning defined?

Even without fully answering this question, let's turn around the conditionals by denoting p and q, respectively, the statements "x is P", and "x is Q". Thus, the first question is how to represent the conditional, or inference, $q < p$, that is, how to link all that with inference. What can be said about the character of an inference once $q < p$ is understood as a statement? Which property, or properties, should a degree of truth t enjoy for saying something, given the truth degree $t(q)$ of the antecedent, on the truth value $t(p)$ of the consequent p?

By presuming that t is such that $q < p \Rightarrow t(q) \leq t(p)$, the degree of true of the conclusion is greater than or equal to that of the antecedent; deduction propagates the truth by nondecreasing it. Note how counterintuitive it would be to suppose $q < p \Rightarrow t(p) \leq t(q)$, propagating the degree of true by retroceding instead of advancing it, something actually odd. What it yet lacks is studying the possibility of having formulae for the degrees $t(q < p)$, depending on $t(q)$ and $t(p)$, as they exist in the classical case in probabilized Boolean algebras by means of the Bayes' formula.

It is interesting, for instance, to bound the truth value of q starting from the modus ponens (MP) "inequality", $q \cdot (q < p) < p$, implying $t(q \cdot (q < p)) \leq t(p)$. Provided it were functions f and g such that,

$$t(q \cdot p) = f(t(q), t(p)), \quad \text{and} \quad t(q < p) = g(t(q), t(p)),$$

the inequality $f(t(q), g(t(q), t(p)) \leq t(p)$ would follow, and perhaps a bounding for $t(p)$ only depending on $t(q)$ could be obtained. For instance, if as in the classical case, it were $f(a, b) = \min(a, b)$, and $g(a, b) = \max(1 - a, b)$, from

$$\min(t(q), \max(1 - t(q), t(p)) \leq t(p) \Leftrightarrow \max(\min(t(q), 1 - t(q)), \min(t(q), t(p))) \leq t(p),$$

the low bounding of $t(p)$: $\min(t(q), 1 - t(q)) \leq t(p)$ would follow.

Those expressions depend on the meaning, in the corresponding language's context, of the linguistic connectives *and, or, not,* whose specification is actually a contextual and partially open problem as formerly said.

In the limit situation in which all the statements are precise, and t can be taken as a Kolmogorov probability, t = prob: Ω à [0, 1], in a Boolean algebra Ω of subsets in the universe X, it appears an interesting case concerning the meaning of the conditional. Provided it were prob(q) > 0, the Bayes' formula prob(p/q) = prob ($q \cdot p$)/prob(q) would represent the probability of $q < p$, of "p if q", a situation in which such a corresponding linguistic conditional cannot be interpreted in Ω as the material form $q' + p$, because it does not give a probability in Ω, as prob(p/q) does.

The conditional, once interpreted by "$p \cdot q$" actually facilitates a probability provided, as in a Boolean algebra, that the conjunction ($\cdot$) were commutative. If prob were a probability in Ω, then prob($./q$) would also be a probability in the set of the traces ($p \cdot q$) all its elements p show in q; that is, prob($./q$) would also be a probability but in the restricted Boolean algebra constituted by the elements in $\Omega^* = \{p^* = p \cdot q; p \text{ in } \Omega\}$. In some sense, q represents a kind of diaphragm only allowing us to consider what "is inside q"; the antecedent constitutes the new universe, hiding what is out of it, and where, for instance, the negation of p is not p', but $p^\wedge = p' \cdot q$. Thus, it is

$$q^\wedge + p* = q' \cdot q + p \cdot q = 0 + p \cdot q = p \cdot q;$$

in Ω^*, inside the new and restricted universe q, but not outside; the Boolean and the conjunctive conditionals do coincide.

All this can suggest a view of measuring conditionals in forms similar to $t(q < p) = t(q \cdot p)/t(q)$, without t being a probability but just a particular measure of "true". Note, for instance, that in the Bayes' formula the conjunction ($p \cdot q$) is supposed commutative because the statements/events are precise/classical sets, but that, when the statements are imprecise/fuzzy sets, such commutative property cannot be generally presumed. The values $t(q)$ and $t(q \cdot p)$ can be called the a priori measures, and $t(q < p) =: t(p/q)$, will be the a posteriori ones.

Provided the conjunction were commutative, it would be easy to find the truth value $t(p < q)$ in function of $t(q < p)$, because $t(q < p). t(p) = t(q \cdot p) = t(p \cdot q) = t(p < q). t(q)$, or

$$t(p<q) = [t(p)/t(q)].\, t(q<p),$$

a formula expressing the truth degree of the inverted conditional, and showing that $t(p < q)$ coincides with $t(q < p)$ if and only if $t(q) = t(p)$. Of course, it also shows that $t(p < q) \leq t(q < p) \Leftrightarrow t(p) \leq t(q)$. The case in which the conjunction is not commutative is an open one.

In some cases, the function $t(p/q) =: t(q \cdot p)/t(q)$ is also a measure in the restricted universe q, inside q, and not only a simple degree of truth of $q < p$. In fact, it is:

1. $p_1 < p_2 = > q \cdot p_1 < q \cdot p_2 = > t(q \cdot p_1) \leq t(q \cdot p_2) = > t(p_1/q) \leq t(p_2/q)$;
2. If p is an a priori working antiprototype, that is, $t(p) = 0$, it is also an a posteriori working prototype, because $q \cdot p < p \Rightarrow t(q \cdot p) \leq t(p) = 0$ implies $t(q \cdot p) = 0$, and $t(p/q) = 0$. Nevertheless, a different problem arises when p is an a priori prototype, because from $t(p) = 1$ it is not immediate to arrive at, or to define, $t(p/q) = 1$, and adding some conditions seems to be necessary for

proving it. For instance, a sufficient condition is the validity, for the pair (q, p), of the law $t(p \cdot q) + t(p + q) = t(p) + t(q)$, because then from $t(p) = 1$ and "$p < p + q \Rightarrow t(p + q) = 1$", it follows that $t(p \cdot q) = t(q)$, and finally $t(p/q) = t(q)/t(q) = 1$. Note that the additive law follows in the particular case with $p \cdot q = 0$; hence such law implies the additive one is more general.

It should be remarked that a priori values are not always attainable from past historical data records; this is the case in some linguistic cases, such as the afore-mentioned "Candidate X will win the election" when, for instance, such a candidate participates in an election for the first time. In such cases, a priori values could be approximate through the experience, or subjective evidence, the observer can (directly or indirectly) have on the corresponding situation; furthermore the contextual knowledge (or evidence) can lead to attributing a value to the a priori ones. Only if there is "contextual ignorance", $t(q) = t(q')$, can it be accepted to take the value $t(q) = 0.5$, because such equality, and provided it were $t(q') = 1 - t(q)$, would imply $t(q) = 1/2$.

If the previous evidence is represented by e, the prior formula can be written in the form

$$t(p/q \cdot e) =: t_e(q < p) = t(p \cdot (q \cdot e))/t(q \cdot e),$$

and only in the ignorance case is it acceptable to take $t(q \cdot e) = 0.5$ that, because from $p \cdot (q \cdot e) < q \cdot e$ it follows that $t(p \cdot q \cdot e) \leq t(q \cdot e) = 0.5$, and the conclusion that follows, $t_e(q < p) \leq 1$, gives no actual information inasmuch as t_e is always less than or equal to one.

For instance, using additional information to specify either a measure m for a predicate P, or a degree t for the validity of a linguistic expression, is nothing more than a priori information (perhaps supplied by an expert, and not always coming from numerical data records; hence, and in a Bayes-style line of thought, it seems necessary for upgrading a priori measures. The use of additional a priori information, or evidence, seems to be a proper and unavoidable resource for the mathematical modeling of commonsense reasoning.

Nevertheless, the measuring of the extent to which a conditional can be a valid one in plain language is indeed an open problem and, when no numerical records are known, it should rely on some subjective information. Note that holding $q \cdot p < p$, and hence $t(q < p) \leq t(p)$, from the former expression follows the upper bound

$$t(q < p) \leq \min(1, t(p)/t(q))$$

for the measure of the conditional.

14.11. For ending this chapter, let's add a comment on conditional statements $p < q$, not from the point of view of deducing, but from that of guessing, and further than was advanced in the former Sects. 5.4 and 11.1. That is, by avoiding deducing from either modus ponens $(p \cdot (p < q)) < q$, or modus tollens (MT)

$(q' \cdot (p < q)) < p'$, and analyzing what can be conjectured or refuted from the sets of premises $\{p, p < q\}$, or $\{q', p < q\}$.

A conjecture and a refutation of the first set of premises (MP) are, respectively, elements c and r, such that $p \cdot (p < q) \leq / c'$, and $p \cdot (p < q) < r'$; concerning the second set of premises (MT), a conjecture is an element d such that $q' \cdot (p < q) \leq / d'$, and a refutation is an s such that $q' \cdot (p < q) < s'$. The problem consists in finding these kinds of four elements.

In principle, it seems easier to confront the refutations, and, concerning the conjectures, those in which it is, respectively, $c' < p \cdot (p < q)$, and $d' < q' \cdot (p < q)$. To "solve" these inequalities, it is necessary to count with a calculus in an algebraic framework, in which both the symbol < and an expression with connectives equivalent to $p < q$ could be identified.

For instance, in the framework of a Boolean algebra, from $q' \cdot (p' + q) \leq r'$, equivalent to $q' \cdot p' \leq r'$, or to $r \leq q + p$, it follows that what refutes a conditional are those r that are below the union of the antecedent and the consequent. Analogously, from $d' \leq q' \cdot (p' + q) \Leftrightarrow d' \leq q' \cdot p' \Leftrightarrow p + q \leq d$ follows that the elements that are greater than the union of antecedent and consequent are conjectures.

14.12. What about the counterfactual conditionals? This is a strongly semantic and scarcely syntactic subject in which background knowledge is, in addition to context and purpose, essential. Without imagining a nonexisting actual situation where the antecedent could be true, the conditional could lose sense.

For instance, and returning to the former example,

- $t(p < q) = \max\ (1 - t(p),\ t(q))$, provided $t(p) = 1$, it would follow that $t(p < q) = t(q)$: If the antecedent is true, the truth value of the conditional coincides with that of the consequent; both should be simultaneously true or false. But, were the antecedent false, $t(p) = 0$, it would follow that $t(p < q) = 1$: a false antecedent produces a true conditional.
- With the conjunctive representation the situation changes for what refers to false antecedents, because $t(p < q) = t(p \cdot q) = \min(t(p), t(q))$, and $t(p) = 1$ implies $t(p < q) = t(q)$, as in the former case, but $t(p) = 0$ implies $t(p < q) = 0$: the conditional only can be true provided its antecedent and consequent were true.

Anyway, with counterfactuals $p < q$ the sources for the truth or falsity of p and q are actually different: p comes from an imaginary source, but q from a real one. It seems that for the analysis of counterfactuals, the usual Boolean model is insufficient, even if the conjunctive one can seem to facilitate a more realistic representation. Of course, there are many more ways of interpreting $p < q$ in Boolean algebras, even without a single expression; for instance, identifying $p < q$ with

- $p' + q$, if p is contradictory with q, and
- $p \cdot q$, if they are not contradictory,

an interpretation satisfying the modus ponens inequality. Namely, in Boolean algebras, the equivalence $p \cdot (p \to q) \le q \Leftrightarrow p \to q \le p' + q$, allows us to construct expressions such as the former easily; for instance,

- q, if $p = 0$, and
- $p \cdot q$, otherwise.

In Boolean algebras there are many possible representations of the conditionals, and finding some of them appropriate to represent a counterfactual conditional is an open topic. For instance, checking if the last form for the conditional, or other similar, can represent some type of counterfactual, could be a way to start such a study, as it also can be by understanding the antecedent as a hypothesis for the consequent.

14.13. Another question with conditionals refers to when a conditional refutes "p and q"; that is, in Boolean algebras, $p \cdot q \le (p \to q)' \Leftrightarrow p \to q \le p' + q'$. In addition, $p < q$ is a type-one speculation of $p \cdot q$, provided $p \cdot q$ were not comparable with $p \to q$, but $(p \cdot q)' \le p \to q \Leftrightarrow p' + q' \le p \to q$. Still, a conditional is a hypothesis for $p \cdot q \Leftrightarrow p \to q \le p \cdot q$. Obviously, for the wild type-two speculations there is no equational way for directly posing and trying to answer the question.

A more realistic and deep study of the conditional's representation problem could come from designing suitable experiments in language, establishing mathematical models and testing them against "reality" and concerning all types of conjectures, doing it, of course, in a not blind but systematic form, within background knowledge, and inside the context-dependent and purpose-driven praxis in plain language and ordinary reasoning.

Nevertheless, a complete study of the symbolic representation and measuring of conditionals still is, in almost all its aspects and for plain language, an open problem affecting, for example, the symbolic representation of children's stories in view of their automatic computer mechanizing, and it is manifested by the more than 40 operators that, in fuzzy logic, have been used to represent conditionals.

References

1. E. Trillas, Some uncertain reflections on uncertainty. Arch. Philos. Hist. Soft. Comput **1**, 1–16 (2013)
2. E. Castiñeira, S. Cubillo, E. Trillas, On possibility and probability in finite Boolean algebras. Soft. Comput **7**(2), 89–96 (2002)
3. L.A. Zadeh, Fuzzy sets as a basis for a theory of possibility. Fuzzy Sets Syst **1**, 3–28 (1978)
4. P. Roeper, H. Leblanc, *Probability Theory and Probability Semantics* (University of Toronto Press, Toronto, 1999)
5. B. De Finetti, Sul significato soggettivo della probabilità. Fundamenta Mathematicae. **XVII**, 298–329 (1931)

6. F.S. Roberts, A note on Fine's axioms for qualitative probability. Ann. Probab **1**(3), 484–487 (1973)
7. E. Trillas, Relational probabilities in intuitionistic lattices. Stochastica **12**(2-3), 85–102 (1988)
8. F. Esteva, On the structure of intuitionistic algebras with relational probabilities. Stochastica **12**(2-3), 103–111 (1988)
9. L.A. Zadeh, The concept of a linguistic variable and its application to approximate reasoning —Part I. Inf. Sci **8**, 199–249 (1975)
10. E. Trillas, L. Eciolaza, *Fuzzy Logic* (Springer, Berlin, 2015)
11. S. Gudder, What is fuzzy probability theory? Found. Phys **30**(10), 1663–1678 (2000)
12. S. Guadarrama, E. Renedo, E. Trillas, *A First Inquiry on Semantic-Based Models of And*, eds. by R. Seising et al. Accuracy and Fuzziness (2015), pp. 331–350
13. L.A. Zadeh, Probability measures of fuzzy events. J. Math. Anal. Appl **23**, 421–427 (1968)
14. P. Weley, G. de Cooman, A behavioural model for linguistic uncertainty. Inf. Sci **134**(1–4), 1–37 (2001)

Chapter 15
Questions on Domesticating and Controlling Analogy

Some additional comments should be added, after the short previous glance at analogy in Chap. 9, that can lead to more questions which still remain open, especially to controlling analogy through degrees of the extent to which it is actually present, and provided such numerical control were possible.

15.1. To begin with, it should be remarked that, in ordinary reasoning, analogy is often expressed in plain language, sometimes with the help of some real or virtual figures; this implies the use, sometimes intensive, of imprecise words in an uncertain setting. Hence, representing analogy would require fuzzy sets, with which if some intervening concepts were precise, they could also be represented by membership discontinuous functions only taking the values 0 or 1.

Both for representing imprecision and controlling uncertainty it could be suitable to count with a "degree of analogy" between pairs of concepts. The values of such degree, varying between 0 and 1, should indicate the total dissimilarity of the two elements if their degree is null, and the total similarity if their degree is one. Provided the degree up to which "x is P is analogous or similar with y is Q" is a number S ($\mu_P(x)$, $\mu_Q(y)$) $\in$ [0, 1], which properties should verify the function S: [0, 1] x [0, 1] $\rightarrow$ [0, 1]?

The properties, $S(a, a) = 1$, and $S(a, N(a)) = 0$, if N is a negation function, undoubtedly seem to reflect properties of analogy; note that the value of S for a pair (μ_P (x), $\mu_P{}^a(x)$) is not necessarily null, but it could be different from 0 because P and P^a can show some degree of analogy.

A typical analogy scheme,

$$a : b :: c : d \Leftrightarrow a \,\text{is to}\, b \,\text{as}\, c \,\text{is to}\, d,$$

should imply $S(a, b) \leq S(c, d)$, although the reciprocal cannot always be sustained, and even if an ε-approximation of the type "it exists a small number $\varepsilon > 0$, such that $IS(a, b) - S(c, d)I \leq \varepsilon$", also could be suitable.

E. Trillas, *On the Logos: A Naïve View on Ordinary Reasoning and Fuzzy Logic*,
Studies in Fuzziness and Soft Computing 354, DOI 10.1007/978-3-319-56053-3_15

What does not seem to be always adequate is presuming $S(a, b) = S(b, a)$ for all a, b in [0, 1], because in general supposing "x is P" is analogous with "y is Q" does not mean that "y is Q" is analogous to "x is P".

To establish a list of axioms for S is not easy; there is no single type of analogy. Anyway, in analogy the problem of breaking it also appears, such as that formerly mentioned with synonyms, and that admits an interpretation by means of the generalized idea on transitivity that follows.

It is said that function S is F-transitive for an operation F: [0, 1] x [0, 1] $\rightarrow$ [0, 1] provided,

$$F(S(a,b),S(b,c)) \leq S(a,c), \text{ for all } a, b, \text{ and } c \text{ in } [0,1].$$

Such definition translates into the degrees, the transitive law "a: b & b: $c \Rightarrow a$: c", allowing a decreasing of degrees in chains of analogy eventually leading to the disappearance of the analogy between the first and last step in the chain. Which properties should be attributed to the operation F?

If it can be accepted that

$$a : b \,\&\, b : c \text{ means the same as } b : c \,\&\, a : b,$$

something not at all odd, the commutative property of F is beyond doubt, as they are beyond doubt that $F(1, 0) = F(0, 1) = 0$, once accepting $S(a, a) = 1$, $S(a, N(a)) = 0$, and F-transitivity; for instance, $F(S(a, a), S(a, N(a)) \leq S(a, N(a))$ just show $F(1, 0) = 0$. It also seems beyond doubt that $F(1, 1) = 1$ once accepting $S(a, a) = 1$. To add $F(0, 0) = 0$ also does not seem rare at all provided it were accepted that "$S(a, b) = S(b, c) = 0 \Rightarrow S(a, c) = 0$". Neither basic properties for S, nor those for F, can be easily fixed; it seems to be something of a contextual character.

In sum, let's call S: [0, 1] x [0, 1] $\rightarrow$ [0, 1], an F-similitude function, or an index of (symmetrical) analogy, provided it were to verify

- $S(a, a) = 1$, for all a in [0, 1],
- $S(a, b) = S(b, a)$, for all a, b in [0, 1], and
- There exists a commutative operation F in [0, 1], verifying the border conditions $F(0, 0) = F(0, 1) = 0$, and $F(1, 1) = 1$, with which it is:

$$F(S(\mu(x),\sigma(x)),S(\sigma(x),\lambda(x))) \leq S(\mu(x),\lambda(x)),$$

for all μ, σ, λ in $[0, 1]^X$, and x in X.

Of course, if more properties were added to S and F, more could follow from this definition. For instance, provided neutrality and monotony, $F(a, 1) = 1$, and $a \leq b \Rightarrow F(a, c) \leq F(b, c)$ for all a, c in [0, 1], respectively, were added to F, then $a \leq 1$ and $c \leq d$ would imply $F(a, c) \leq F(1, d) = d$. Thus, provided it were $\varepsilon \leq a$, and $\varepsilon \leq b$, it would be

$$F(\varepsilon, \varepsilon) \leq F(a, b), \text{but not necessarily } \varepsilon \leq F(a, b),$$

unless it were $\varepsilon \leq F(\varepsilon, \varepsilon)$. Note that, although with some reservation for what concerns symmetry in analogy, it is presumed that S is symmetrical, and it means that the former definition of S cannot be considered too general, but only suitable for a symmetrical analogy.

Let's show an example with $F = W_f$, a t-norm in the Lukasiewicz family that, obviously, can be taken as a function F. If it is

$$\varepsilon \leq S(\mu(x), \sigma(x)), \quad \text{and} \quad \varepsilon \leq S(\sigma(x), \lambda(x)),$$

it follows that

$$W_f(\varepsilon, \varepsilon) = f^{-1}(\max(0, 2f(\varepsilon) - 1) \leq W_f(S(\mu(x), \sigma(x), S(\sigma(x), \lambda(x))) \leq S(\mu(x), \lambda(x)))$$
$$\Rightarrow \max(0, 2f(\varepsilon) - 1) \leq f(S(\mu(x), \lambda(x)))).$$

That is, provided $2f(\varepsilon) - 1 \leq 0 \Leftrightarrow f(\varepsilon) \leq 1/2 \Leftrightarrow \varepsilon \leq f^{-1}(1/2)$, nothing could be concluded, and only if $1/2 < f(\varepsilon) \Leftrightarrow f^{-1}(1/2) < \varepsilon$, would it follow that $0 < f^{-1}(2f(\varepsilon) - 1) \leq S(\mu(x), \lambda(x))$. Hence, under the hypotheses that $\mu(x)$ corresponds to "x is P", $\sigma(x)$ to "x is Q", $\lambda(x)$ to "x is R", and that it is $f^{-1}(1/2) < \varepsilon$, it can be established that if "x is P" is similar to "x is Q", and this statement is similar to "x is R", with respective degrees greater than ε, then the first statement is similar to the third with a degree greater than $f^{-1}(2f(\varepsilon) - 1)) > 0$.

To keep $\varepsilon \leq S(\mu(x), \lambda(x))$, it suffices that

$$\varepsilon \leq f^{-1}(2f(\varepsilon) - 1) \Leftrightarrow 1 \leq f(\varepsilon) \Leftrightarrow 1 \leq \varepsilon \text{ implying } \varepsilon = 1,$$

and the three statements are fully similar. Hence, if, as usual, $0 < \varepsilon < 1$, it would be $0 < f^{-1}(2f(\varepsilon) - 1) < \varepsilon$, showing that the degree of similarity between the first and third statements actually decreased; consequently, where $\varepsilon \in (0, 1)$, a threshold of analogy, that is, a level under which analogy ceases to be preserved, the analogy would be lost by not surpassing such a threshold. It can happen in three steps as in the example, or in more steps, but the decreasing values show that arriving at a step not keeping analogy with the first can be assured.

What happens with F = prod? With F = prod, $\varepsilon^2 \leq S(\mu(x), \lambda(x))$ is obtained, and because for $0 < \varepsilon < 1$ it is $0 < \varepsilon^2 < \varepsilon$, the same conclusion follows.

For not having decreasing degrees, it should be $F(\varepsilon, \varepsilon) = \varepsilon$, and this is only possible (with continuous *t*-norms) either if F = min, or F is an ordinal sum counting ε among its idempotent elements. Anyway, with these last *t*-norms the chains of analogous statements will not break, and this seems to be something actually rare in analogy, as it is with synonymy that can be seen as a linguistic phenomenon of graded analogy of meaning. Among *t*-norms only those in the family of W seem suitable for modeling analogy's breaks.

15.2. In artificial intelligence, and namely in case-based reasoning where reasoning is conducted by analogy, it is important to control analogy by means of a threshold. For instance, there is a system trying to mechanize geometrical reasoning by analogy, in which the function S is given by:

$$S(a,b) = \sum \min(a_i, b_i)/\max\left(\sum a_i, \sum b_i\right) \in [0,1],$$

provided both figures a and b were characterized by the same attributes A_1, …, A_n, but satisfying each one with degrees a_i and b_i in [0, 1] ($1 \leq i \leq n$), respectively. As it is easy to see, it is,

- $S(a,b) = 1 \Leftrightarrow a_i = b_i$ for all $i \in \{1,2,\ldots,n\}$;
- $S(a,b) = 0 \Leftrightarrow \min(a_i,b_i) = 0$, for all I;
- Provided S were F-transitive, $F(a, b) = 0$ should imply $a + b \leq 1$, and, provided it were $F \leq W$, then S is F-transitive.

Hence, by constraining F to be a continuous t-norm, it should be a t-norm W_f, smaller than W, and neither min, nor prod, make SF-transitive.

In praxis, by trial and error, a threshold of analogy was empirically found equal to 0.7 with which the system works well. Note that for approaching the former low bound $0 < f^{-1}(2f(\varepsilon) - 1)$ to 0.7, it suffices to take the order automorphism $f(x) = x^2$, with which it is $W_f \leq W$, and

$$\max\left(0, 2\varepsilon^2 - 1\right)^{1/2} > 0 \Leftrightarrow \sqrt{(1/2)} < \varepsilon.$$

Because $\sqrt{(1/2)} \sim 0.707107$, it suffices to take $0.708 \leq \varepsilon$ as the threshold; as the builders of the system only appreciate up to the first decimal digit, they just took 0.7 instead of 0.708. W_f-transitivity helps to foresee a lower bound for the threshold of analogy.

Formulae such as the former are important for the goal of "controlling" analogy. In fact, analogy often depends on the attributes that are taken for considering it; for instance, as said, if the only considered attribute is "spherical form", oranges and apples could be seen as similar, but as soon as more attributes such as color, taste, smell, and so on, are considered, such similarity would cease. Analogy is often used, as it is in case-based reasoning, to substitute an element a by another b seen as similar, and on which is known more than is known on a; for instance, in the former example and when $S(a, b) > 0.7$, the properties shown by figure a are presumed for figure b. This permits, for reasoning on b, to guess that it enjoys the properties of a which are under consideration.

The importance of controlling analogy directly comes from the necessity of not seeing "a like b" when the degree of analogy between a and b is too low, in short and joking, for not eating apples instead of oranges, or oranges instead of apples. It is for this reason that, once an index of similarity or analogy S is known, it is important to fix a threshold of analogy; only two objects a, b such that the index $S(a, b)$ is greater than the threshold, could be interchanged for reasoning. On the contrary, not doing it is foolish and can actually lead to (saying it metaphorically) confounding melons and footballs.

Hence, analogy deserves to be controlled and, in such respect, it is an open problem to know which indexes of analogy are, actually, measures of the contextual meaning of the word "analogous", for which the empirical relationship "less analogous than" should be previously captured.

15.3. Reasoning by using analogy for conjecturing or for refuting is not always controlled by means of numerical indexes such as, for instance, is not done in commonsense reasoning where numerical comparisons are not usual. Anyway, it does not exclude some kind of verbal or linguistic control by, for instance, attending to the diverse attributes in play and limiting the analogy to them; it is the case when saying that John and his sister Anne have the "same eyes as their mother". In these cases, it is obvious that it is not said that John and Anne are fully similar, but just that they are physically similar with respect to the attribute "eyes"; they hold no similarity with respect to mouth, ears, hair color, hands, and so on, and, provided the number of these attributes were 10, it could be said that they have $1/10 = 0.1$ as a similarity index, *a* value that is very far from the value 1 of full similarity. It is more difficult to establish an index when considering not-physical characteristics of John and Anne, such as social behavior, gesticulating, and the like, or something else that can also be appreciated such as, "John and Anne argue like their parents did." A very simple index of analogy is the ratio "positive attributes/total number of attributes", that, nevertheless, does not take into account the degrees to which attributes are satisfied. Anyway, analogy should be always controlled; numerically if possible.

Analogy is still full of mystery concerning the establishment of a suitable mathematical framework for its formal study, that is, for its scientific domestication; but what is beyond doubt is that the analogy between two objects, images, persons, and the like, is rarely considered an identity even if identity can be seen as a particular case of analogy. Analogy always refers to some peculiarities of what is considered, but not always to its "totality", something that often enough is only metaphysical. For instance, the John that currently is 65 years old, appearing in a photo of him taken 50 years ago, is different but analogous, similar, to the current John, this to the extreme of saying that both are the same John. Not all characteristics of the second John are taken into account, but only some of them.

Also for instance, in the framework of the curves given by quadratic equations, it can be said that a circle and an ellipse whose focal points are close are similar figures, even if they are not identical and nobody with some elemental knowledge of mathematics will confuse circles with ellipses, such as balls with melons; in the end, a circle is but a particular case of ellipse when the two focal points coincide. Anyway, in a problem of graphical design, it is perfectly conceivable to see both figures as analogous with the aim of graphically translating to the circle some property of the ellipse; even in plane geometry there are problems whose solutions are well conjectured after a graphical approximate reasoning with analogous figures.

15.4. A short additional reflection is still in order. In the same vein that circles and ellipses cannot be confused, as well as the John of 50 years ago is not physically the same as today's John, for a rough reasoning in commonsense reasoning, analogy usually forgives some attributes and just concentrates on others, but, for doing reasonings such as those of science, analogy should be refined.

Beyond formal frames, identity is often a kind of illusion; there is no more identity than a full coincidence in all the imaginable orders, that is, a is identical to b if and only if a were to coincide fully with b, that is, if a could be substituted by b everyplace; it is the old Leibniz "identity of indiscernibles", when two real objects in the world, or two virtual objects in the brain, are not identical, are just different, and analogy is but a form of graduating differences in such a way that when the degree is one are taken as identical, and only are fully different when the degree is zero. In this sense, the symbols $=$ and $\neq$ are but particular exceptions of the symbol $\sim$ denoting analogy, and the human senses seem to count with an innate possibility for not only grasping difference, but refining it to analogy.

Surely, without such intellectual capability, the evolution of mankind would be another and different. This capability is possibly already inserted in the brain, with repercussions on reasoning, where analogy should be seen, as said before, as something essential, for instance, for passing from the perception of "big tree", to "big mountain", to "big number", and to "big money". From perceiving that stones are of different sizes, to see that different stones can be (imprecisely, of course) classified in several subtypes, such as small, big, and middle, only with time, and when the ideas of weighting and its measuring appeared and numerical degrees established, did the idea of numerical degree of similitude appear.

15.5. To end this chapter, let's remark that analogy between things, or between physical situations, passed to analogy between virtual objects and situations, passed to fiction, and to metaphorical stories such as those for children and, finally, to intellectual tasks such as writing novels, philosophy, and also to scientific research. The case of the falling lift of Einstein, the falling apple of Newton, or the ouroboros of Kekulé, are but examples of analogies suggesting a conjecture that, in science, is not only for telling it, but that it should, first of all, be proven by some reasoning perhaps supported by an F-similitude function.

In the same vein, when the classical logical calculus was generalized to the fuzzy one, and the universal connectives *and*, *or*, *not*, and *if/then*, passed to operations that should be specified at each particular context, the difference suggested that, in large statements, or in systems of many rules, linguistic connectives expressed by the same word deserve to be specified by different operations at each of the statement parts depending on their respective meanings. For instance, in the statement "John and Anne bought a new house, and thought of either repainting all of it, or modernizing the kitchen," the first "and" seems to be commutative, but the second seems to be not; hence, if translating it into fuzzy terms, the first could be represented by a t-norm, but to represent the second a noncommutative operation should be chosen.

15.6. Another analogy that can give birth to fuzzy sets came from comparing the $\{0, 1\}$-valued characteristic function of a crisp set with the truth values of precise statements and then, by analogy, enlarging this set of values to the full unit interval [0, 1] and allowing the "new sets" corresponding to imprecise statements to be characterized by a function ranging in the unit interval. This is a view not supposing that fuzzy sets are a simple generalization of crisp sets, but obtained by some kind of aggregation of them. It is a presumption that, by enlarging the analogy, does not lead to considering only the basic connectives min, max, and $1 - \text{id}$, in particular to believe that "p and p", and "p or p" always mean p.

For instance, if, in the universe $X = \{1, 2, 3, 4, 5\}$, the sets $A = \{1, 2, 3\}$ and $B = \{3, 4, 5\}$, are "operated" (through their respective characteristic functions) with the arithmetic mean $M(x, y) = (x + y)/2$, what results is

$$1/0.5 + 2/0.5 + 3/1 + 4/0.5 + 5/0.5$$

that, represented by $M(A, B) = (A + B)/2$, could suggest, by analogy, to consider all the entities obtained by applying aggregation functions to sets, such as all the means, as fuzzy sets. Also, for instance, the noncommutative pondered mean $N(x, y) = (x + 3y)/4$ leads to

$$N(A,B) = (A+3B)/4 = 1/0.25 + 2/0.25 + 3/1 + 4/0.75 + 5/0.75.$$

These examples serve to pose the question of which membership functions can come from aggregating a finite number of crisp sets, and to observe that commutativity cannot always hold with them because, for instance, it is $M(A, B) = M(B, A)$, but $N(A, B) \neq N(B, A)$.

Although not all fuzzy sets can come from aggregating a finite number of crisp sets, it does not mean that given a fuzzy set μ on X there is not a family of aggregation functions $\{A_x; x \in X\}$, each giving a value of the membership function $\mu(x) = A_x(Ch(x))$ for some numerical characteristic $Ch(x)$ of each x such as in the limiting case of a single aggregation, with the formers $M(A, B)$, and $N(A, B)$. At the end, it is similar to what was formerly said on obtaining the numerical values $\mu(x)$ by a statistical methodology.

References

1. E. Castiñeira, S. Cubillo, E. Trillas, On a Similarity Ratio; Proceedings. EUSFLAT-ESTYLF (1999), pp. 239–248
2. F. Chouraqui, C. Inghilterra, A Model of Case-Based Reasoning for Solving Problems of Geometry in a Tutorial System, in 'Intelligent Learning Environments: The Case of Geometry'. ed. by J.-M. Laborde (Springer, Berlin, 1996), pp. 1–16
3. B. Bouchon-Meunier, L. Valverde, A Resemblance Approach to Analogical Reasoning Functions, in 'Fuzzy Logic for Artificial Intelligence'. eds. by J.P. Martin et al. (Springer, Berlin, 1997), pp. 266–272
4. F. Klawonn, J.L. Castro, Similarity in fuzzy reasoning. Mathware Soft Comput **2**, 192–228 (1995)

Chapter 16
Questions on the Classical Schemes of Inference

In the classical calculus with precise concepts, some schemes of deductive inference are used such as the modus ponens (MP), and the modus tollens (MT), but also the so-called disjunctive mode, among others. They are instances of what is often known as the Aristotelian logical forms, and are of some interest for the mechanizing of formal deduction.

In what follows a scrutiny of the validity of these schemes is conducted for first certifying them in Boolean algebras, second (in Chap. 5) to know which conditions can hold in a basic fuzzy algebra (BAF), and finally what can be said about their validity in the general case of ordinary reasoning under the model of natural inference. Of course, several laws should be applied for proving such schemes in the classical calculus, laws that, in general, cannot always be presumed in ordinary reasoning.

16.1. Concerning the schemes of modus ponens and modus tollens, respectively,

$$p, p < q : q, \quad \text{and} \quad q', p < q : p',$$

they can be posed in three not properly coincidental forms; the first is purely algebraic, the second is tautological, and the third concerns truth values for modus ponens. They are the following.

- p and $p \to q = p' + q$, imply q;
- $p = 1$, and $p \to q = p' + q = 1$, imply $q = 1$,
  and
- $t(p) = t(p \to q) = t(p' + q) = 1$, imply $t(q) = 1$.

The first is formally proven by: $p \cdot (p' + q) = p \cdot q \leq q$; the second follows immediately because $p = 1$, means $1 = p' + q = 0 + q = q$; and, concerning the third $t(p) = 1$ implies $1 = t(p' + q) = \max(1 - t(p), t(q)) = \max(0, t(q)) = t(q)$.

E. Trillas, *On the Logos: A Naïve View on Ordinary Reasoning and Fuzzy Logic*, Studies in Fuzziness and Soft Computing 354, DOI 10.1007/978-3-319-56053-3_16

Hence, the Boolean model actually certifies that the three versions of the scheme hold.

It should be noted that MP holds universally in the former model of ordinary reasoning, provided the "conditional statement" $p < q$, once p is known, would effectively allow a "movement" up to q. Concerning MT, once q' is known, and because $p < q$ implies $q' < p'$, if p' can be effectively reached, MT also holds in the general model. In it, MT is a consequence of MP (q': $p < q$ => q': $q' < p'$:: q'), although the reciprocal cannot always be stated.

Regarding the Boolean case, the three previous forms also hold for MT because $p \rightarrow q = q' \rightarrow p'$; MT is only equivalent to MP in the Boolean framework.

Regarding the Boolean case with $p \rightarrow q$ expressed in conjunctive form $p \cdot$ q, note that MP always holds in all lattices because $(p \cdot (p \rightarrow q)) = p \cdot (p \cdot q) = p \cdot q \leq q)$, but MT cannot hold in Boolean algebras because $q' \cdot (p \cdot q) = 0$. It should be pointed out that, were the lattice a De Morgan algebra, then $q' \cdot (p \cdot q) = p \cdot (q \cdot q')$ would not always be 0, hence $q' \cdot (p \rightarrow q) = q' \cdot (p \cdot q) = p \cdot (q \cdot q') = q \cdot (p \cdot q') \leq q$, and MT holds for the non-Boolean elements. Thus, the validity of MT not only depends on the laws of the corresponding algebraic structure, but also in how the conditional $p \rightarrow q$ is expressed.

16.2. Consider the disjunctive scheme:

$$\text{If} \quad p' \quad \text{and} \quad p+q, \text{then} \quad q.$$

In fact, it follows from $p' \cdot (p + q) = p' \cdot q \leq q$, and q can be concluded. Another form of posing this scheme is:

$$p' = 1, \quad p+q = 1, \quad q = 1$$

by presuming that both p' and $p + q$ are tautologies. Thus, $p = 0$ and $1 = p + q = 0 + q = q$.

Finally, the last form is with just truth values:

$$t(p') = 1, \quad t(p+q) = 1, \quad t(q) = 1$$

where neither p', nor $p + q$, are necessarily tautologies but have truth value one. In this case, $t(p') = 1 - t(p)$ implies $t(p) = 0$ and $1 = t(p + q) = \max(t(p), t(q)) = t(q)$. Thus, $t(q) = 1$ is concluded.

Under the three forms of posing the question, q is concluded, and the Boolean model certifies a scheme that has been accepted from very old times in precise deductive reasoning.

Concerning its possible universal validity in the general model, such as those of the schemes MP, MT, once $q < p + q$ is accepted, it follows that $(p + q)' < q'$; thus once q' is obtained and provided the negation of q were intuitionistic, $(q')' < q$ (or strong), q is forward reached from the negation of q'. Hence, with the conditions that the negation reverses the inferential relation < and it is intuitionistic or strong in particular, the disjunctive scheme holds universally.

16.3. Regarding the scheme of proving by reduction to absurdity,

$$q, p < q' : p',$$

expressed in its three versions in a Boolean algebra,

- $q \cdot (p' + q') = q \cdot p' \leq p'$;
- $q = 1,\ 1 = p' + q' = p' + 0 = p'$, and
- $1 = t(q) = t(p' + q') = \max(t(p'), 1 - t(q)) = \max(t(p'), 0) = t(p')$,

is also certified in the Boolean model. Note that it does not hold in a lattice whatsoever with $p \rightarrow q = p \cdot q$, inasmuch as $q \cdot (p \rightarrow q') = q \cdot (p \cdot q') = 0$.

Regarding the general model, provided the negation of q were weak, or strong in particular, that is, it would verify $q < (q')'$, because from $p < q'$ follows $(q')' < p'$; then and provided the negation reverses <, and the triplet $(q, (q')', p')$ is transitive, it would result in $q < p'$. Thus, under the conditions of reversing <, weak or strong negation, and transitivity, *reductio ad absurdum*, holds universally; but, provided one of these conditions were to fail, the question would remain open. Reduction to absurdity is risky when <- transitivity fails.

16.4. The so-called scheme of resolution,

$$p' < q, q < s : p + s,$$

algebraically follows from

$$(p + q) \cdot (q' + s) = p \cdot q' + p \cdot s + q \cdot s \leq p + s + s = p + s,$$

with the distributive law playing a pivotal role.

In the case of tautologies, $p + q = q' + s = 1$, it also follows that $1 \cdot 1 = 1$ $p + s$, or $p + s = 1$.

With truth values, $t(p + q) = t(q' + s) = 1$, or $\max(t(p), t(q)) = 1$, and $\max(1 - t(q), t(s)) = 1$. The first implies either $t(p) = 1$, or $t(q) = 1$. If $t(q) = 1$, the second shows $t(s) = 1$; if $t(p) = 1$, then $t(q)$ can be either 0 or 1, and if $t(q) = 0$, $t(s)$ is whatever 0 or 1. In conclusion $t(p + s) = \max(t(p), t(s))$ always equals 1.

Note that this scheme holds in all lattices when $p \rightarrow q = p \cdot q$, because $(p' \rightarrow q) \cdot (q \rightarrow s) = p' \cdot q \cdot s \leq s \leq p + s$.

Concerning its universal validity in the general model, note that if the triplet (p', q, s) is transitive, it follows $p' < s$, and once $s < p + s$ is accepted, provided the triplet $(p', s, p + s)$ were also transitive, it would be concluded that $p + s$. Hence, the scheme of resolution could be stated in plain reasoning under the transitive law for <, but not without it.

16.5. The "constructive dilemma" is the scheme,

$$p + q, p < r, q < s : r + s,$$

that, in the case of tautologies and because it is $p \leq q \Leftrightarrow p' + q = 1$, follows immediately from $p \leq r$, $q \leq s \Rightarrow p + q \leq r + s$.

Algebraically, it also follows from

$$(p + q) \cdot (p' + r) \cdot (q' + s) = p \cdot q' \cdot r + p \cdot r \cdot s + q \cdot p' \cdot s + q \cdot r \cdot s$$
$$\leq r + r + s + s = r + s.$$

With truth values,

$t(p + q) = t(p' + r) = t(q' + s) = 1$, and,

- if $t(p) = 1$, follows $t(r) = 1$;
- if $t(p) = 0$, follows $t(s) = 1$;

thus, in both cases, it is $t(r + s) = \max(t(r), t(s)) = 1$.

Concerning its universal validity in the general model, once transitivity is presumed, it is $p < r + s$ and $q < r + s$, and once it is accepted that $p < p + q$ ($q < p + q$), it is a backward path from $p + q$ to $p(q)$, and a forward one from p (q) to $r + s$. Hence under transitivity, $r + s$ can be reached from $p + q$ with combined forward–backward "movements". In plain reasoning the constructive dilemma can fail if transitivity fails.

16.6. The scheme

$$p' + q', r < p, s < q : r' + s',$$

is called the "destructive dilemma", and the proofs of it in a Boolean algebra are obtained analogously to those of the constructive dilemma.

Concerning its universal validity, because it follows that $p' < r'$ and $q' < r'$, it also follows that $p' < r' + s'$ and $q' < r' + s'$, provided the transitive law were to hold. Hence, from $p' < p' + q'$ ($q' < p' + q'$), it is a backward path from $p' + q'$ up to $p'(q')$, and a forward path from $p'(q')$ up to $r' + s'$, that allows reaching $r' + s'$. In any case, $r' + s'$ can be reached from $p' + q'$. Under transitivity, it holds under a combined backward-forward movement, but it can fail with a lack of transitivity.

16.7. The four most well-known schemes of deductive reasoning, the ancient "modus" of the old logic, are the following.

- MP, Modus ponendo ponens, shortened to modus ponens
- MT, Modus tollendo tollens, shortened to modus tollens
- MPT, Modus ponendo tollens, shortened to disjunctive scheme,
- MTP, Modus tollendo ponens, $(p \cdot q)', q : p'$.

Of them, it only lacks reviewing MTP in Boolean algebras, and reflecting on the possibility of its universal validity. The first is immediate after proving what seems was unknown in Middle Ages logic: That the four "modi" are equivalent in a Boolean framework to $p \to q = p' + q$. Such proof is as follows.

- $p \to q = q' \to p'$, shows MP $\Leftrightarrow$ MT.
- $p + q = p' \to q$, shows MT $\Leftrightarrow$ MPT
- $(p \cdot q)' = p' + q' = q \to p'$, shows MP $\Leftrightarrow$ MTP.

Hence, although MTP can be directly proven algebraically by the Boolean calculation $(p \cdot q)' \cdot q = (p' + q') \cdot q = p' \cdot q \leq p'$, its equivalence with MP allows avoiding any additional consideration.

Regarding MTP's possible universal validity in the general model, it is obvious that it cannot be certified by the last considerations, and, in particular, due to the Boolean identification of $p < q$ with $p' + q$. Nevertheless, as shown in the dilemmas, the following can just be said. Because $p \cdot q < p$ implies $p' < (p \cdot q)'$, provided it were accepted that this also implies $q \cdot p' < q \cdot (p \cdot q)'$, it would then be clear that, once the data $q \cdot (p \cdot q)'$ are known, $q \cdot p'$ can be backward reached, and that from $q \cdot p' < p'$, p' is finally forward reached. Hence, provided the law of monotony $a < b \Rightarrow c \cdot a < c \cdot b$ for any c, were accepted, p' could be concluded after a backward deduction up to $q \cdot p'$, and a forward one up to p'. In any case, such a law of monotony does not seem to be a bizarre one for ordinary reasoning or, at least, for some parts of it, and what remains an open question is to know on which weak suppositions the four modi can be equivalent in the general model of ordinary reasoning.

Let's recall that the Latin words *ponendo* and *tollens*, mean "placing" and "suppressing," respectively. For instance, in the MTP it refers to reaching truth by first suppressing it, and secondly, placing it; in MPP = MP, it refers to reaching truth by first placing it, and secondly also placing it. In the naïve symbolic representation managed here, placing corresponds to doing a forward movement, and suppressing to a backward one. The old terminology still keeps some significance, and it is credible that in Middle Age's scholastics the modi were seen in a form that the new general model reproduces in different and symbolic terms.

In the old scholastic logic, the four modi were not seen as equivalent. Because these modi are not known to be equivalent for all kinds of ordinary reasoning, neither Boolean algebras, nor non-Boolean De Morgan algebras, nor non-Boolean orthomodular lattices, nor BAF, would be possible models for the totality of ordinary reasoning, but only for some and perhaps very small parts of it. It is clear that a more general mathematical framework is necessary for a formal global study of ordinary reasoning.

16.8. What has been presented in this section only refers to deduction, but what about conjecturing and refuting by means of the classical schemes? Something was further advanced on MP and MT, but it still lacks asking for the other schemes. Is there some actual possibility for using them to refute or to conjecture? For instance,

MT and MTP are actually ways for deductively refuting p, and MP and MPT for deductively proving p. What can be said about conjecturing or refuting with the schemes of resolution and MPT?

For instance, can the data in MPT serve p' and $p + q$ to refute q? In the Boolean framework it is equivalent to satisfy the inequality $p' \cdot (p + q) = p' \cdot q \leq q'$, implying $p' \cdot q = 0$, forcing $q = q \cdot p + q \cdot p' = q \cdot p$; that is, $q \leq p$ or $q' + p = 1$: $q \rightarrow p$ should be a tautology.

Can these data serve to reach q as a type-one speculation? This is equivalent to both $p' \cdot (p + q)$ NC q, and $q' \leq p' \cdot (p + q)$, but the first is impossible because it is $p' \cdot (p + q) = p' \cdot q \leq q$, and the second would imply $q' \leq p' \cdot q$, meaning $q' = 0$, or $q = 1$. No type-one speculation is available from $\{p', p + q\}$.

Can q be a hypothesis? Can it be $q \leq p' \cdot (p + q) = p' \cdot q$? It would imply $q = p' \cdot q \Leftrightarrow q \leq p'$ that q is contradictory to p and that p already refutes q.

The scheme of reduction to absurdity actually refutes p. Can it serve to say that p' is a hypothesis for the data q and $p < q'$? It would require $p' \leq q \cdot (p' + q') = q \cdot p' \Leftrightarrow p' = \mathrm{q} \cdot p' \Leftrightarrow p' \leq q \Leftrightarrow p' \leq (q')'$ that p' and q' should be contradictory and that q' already refutes p'. For serving to conjecture p, it should be $p' \cdot q$ NC p, something that is actually possible provided $p \neq 0$ and $p' \cdot q \neq 0$, because if it were $p \leq p' \cdot q$ it would imply $p = 0$, and if it were $p' \cdot q \leq \mathrm{p}$ it would imply $p' \cdot q = 0$. Thus, for being a type-one speculation it should be $p' \leq p' \cdot q \Leftrightarrow p' = p' \cdot q \Leftrightarrow p' \leq q \Leftrightarrow q' \leq (p')'$, q' and p' should be contradictory; p' should already refute q'.

Can the scheme of resolution, whose data are $p' < q$ and $q < s$ serve for refuting $p + s$? It should be $(p + q) \cdot (q' + s) = p \cdot q' + p \cdot s + q \cdot s \leq (p + s)' = p' \cdot s'$, implying $p + s = 1$.

- Can it serve for conjecturing $p + s$? The answer is no, because the former expression is, obviously, less than or equal to $p + s$.
- Can $p + s$ be a hypothesis? For it, the former union should equal $p + s$: $p \cdot q' + (p + q) \cdot s = p + s$, a Boolean equation whose solution could enlighten a possible answer. For instance, from it follows $p \cdot q' \cdot s' = p \cdot s' \Leftrightarrow p \cdot s' \leq q' \Leftrightarrow q \leq p' + s$, meaning that the possible involved triplets (p, q, s) should be searched between those verifying $q \leq p' + s$, and so on.

16.9. Although the typical scholarly proofs by means of truth-tables hides them, the rules for deducing consequences can be translated into the Boolean formal model by means of equations and inequalities. It also happens with hypotheses and, in some cases, with type-one speculations, but never with type-two speculations just characterized by the lack of comparability with the reasoning's premises and its negation.

This does not mean, nevertheless, that these speculations are never accessible step by step through some forward and backward deductive paths; perhaps some of those speculations could be reached by means of a system of inequalities mixing forward ($\leq$), and backward ($\geq$) paths.

For instance, in a finite Boolean algebra with five atoms a, b, c, d, e, taking $a + c$ as the résumé of the premises, $b + d$ is neither below nor after $a + c$, and is $(a + c)$ NC $(b + d)$, and $(a + c)' = b + d + e \geq b + d$; hence $b + d$ is a type-one speculation of $a + c$. Nevertheless, because $a + c \leq a + c + b + d \geq b + d$, $b + d$ is reachable from $a + c$ by means of a first forward movement up to $a + c + b + d$, followed by a backward one to $b + d$. It analogously happens with $a + c$ and $b + c$, for which it is $(a + c)$ NC $(b + c)$, $a + c$ NC $(b + c)' = a + d + e$, and hence $b + c$ is a type-two speculation of $a + c$; but because $c \leq a + c$, $c \leq b + c$, the speculation is reached by a backward movement up to c followed by a forward one up to $b + c$.

Characterizing speculations that can be reached by a sequence of backward and forward movements, that is, that are algorithmically reachable step by step, is an open question surely dependent on the formal framework; anyway, those that are not reachable are the properly inductive or creative speculations.

Another open topic, perhaps related to this last, refers to obtaining a definition of the heuristics used in artificial intelligence programs, and for which it is needed to know something previously on a searched conclusion.

Because (deductive) logic cannot be seen as a subject only concerning the preservation of Aristotelian logical forms, nor as only doing reasoning by using them, it is important to study whether there are other forms that, even possibly of an approximate character, can be useful for not only doing deductive reasoning. Its existence can be, eventually, of relevance for the computer mechanization of ordinary reasoning.

References

1. E. Trillas, C. Alsina, E. Renedo, On some schemes of reasoning in fuzzy logic. New Math. Nat. Comput **7**(3), 433–451 (2011)
2. F. Bex, Ch. Reed, Schemes of inference, conflict, and preference in a computational model of argument. Stud. Logic Grammar Rhetoric **27**(36), 39–53 (2011)
3. Sister Miriam Joseph, *The Trivium: The Liberal Arts of Logic, Grammar, and Rhetoric* (Paul Dry Books, Philadelphia, 2002)

Chapter 17
Questions on Fuzzy Inference Schemes

As soon as the field of precise words is abandoned, the topic of the Aristotelian logical forms, the deductive inference schemes, changes; when imprecision and uncertainty appear, everything should be designed and approximation present. In any case, a formal framework, that is, a particular basic fuzzy algebra (BAF), should be found for knowing if a given scheme holds in it; as always, the schemes do not hold at any fuzzy algebra, but they can hold in some of them. Only after knowing which can be such formal frameworks, can the corresponding scheme be freely used for conducting a formal reasoning within it; in a given fuzzy framework in which a scheme can be freely used, there may be other schemes that cannot.

With imprecise words, the classical schemes of deductive reasoning do not hold in a unique framework. Hence, their use in a representation of ordinary reasoning is not always fully guaranteed, but bounded to some mathematical frameworks, that is, within some BAFs.

In any case, the fact that some of the before-considered schemes can hold (in one way or another, but with transitivity) in the new and general frame of ordinary reasoning with the primitive relation $<$, indicates that for each of them a suitable fuzzy algebra, or a family of them, in which the scheme holds, can exist; of course, what is not sure is that such an algebra is functionally expressible by a standard continuous triplet (T, S, N) allowing us to pose it by an inequality and to be solved thanks to the properties such functions enjoy.

To overview this topic, let's presume that the appearing membership functions are designed to represent some linguistic labels in a universe of discourse X. In what follows, everything is analyzed in a standard algebra $([0, 1]^X, T, S, N)$, because it is possible to solve functional equations and inequalities in them, although it is not always easy, thanks to the many laws of continuous t-norms, t-conorms, and strong negations. Thus, the question is the mathematical one concerning whether the schemes hold inside some of these particular BAFs.

E. Trillas, *On the Logos: A Naïve View on Ordinary Reasoning and Fuzzy Logic*,
Studies in Fuzziness and Soft Computing 354, DOI 10.1007/978-3-319-56053-3_17

17.1. Let's first do a review of the modus tollens (MT) scheme:

$$\sigma', \mu \to \sigma : \mu',$$

that, translated into a standard BAF, and as does the formerly seen modus ponens (MP) scheme, gives the functional inequality

$$T\,(N\,(b),\,J\,(a,b)) \leq N(a), [*],$$

for all a, b in [0, 1] and the unknowns T, N, and J. Hence, the MT inequality is equivalent to

$$J(a,b) \leq J_{\mathrm{T}}(N(b),N(a)), \quad \text{with} \quad J_{\mathrm{T}}(a,b) = \mathrm{Sup}\{z \in [0,1] : T(z,a) \leq b\},$$

not only showing that $J_{\mathrm{T}} \circ (N \times N)$ is the greatest function with which the MT inequality holds once a continuous t-norm T and a strong negation N are chosen, but that for any continuous pair (T, N), J verifies inequality [*] if and only if it is $J \leq J_{\mathrm{T}} \circ (N \times N)$.

For instance, and as is easy to check, $J(a, b) = a.b$ (typical of fuzzy control) neither verifies the MT inequality with T = prod, nor with T = min, and $N = 1 - \mathrm{id}$. For its part, and being $W(1 - b, a.b) = \max(0, b(a - 1)) = 0$, because $a \leq 1$, meaning that the conjunction of the premises, $\sigma' \cdot (\mu \to \sigma)$ is μ_0, the membership function of the empty set, the function J = prod, verifying the MP inequality with any t-norm (because J verifies it with min), cannot be used to represent a conditional that should verify the deductive rule of MT.

Instead, the function $J(a, b) = \max(1 - a, b)$, verifying MP with $T = W$, also verifies MT with this t-norm, because it is contra symmetrical: $J(1 - b, 1 - a) = \max(1 - (1 - b), 1 - a) = \max(b, 1 - a) = \max(1 - a, b) = J(a, b)$. In addition, the W-conjunction of the premises, $W(1 - b, \max(1 - a, b)) = \max(0, 1 - (a + b))$ is not constantly null; hence this J can be safely used for MT.

For using a scheme of reasoning, it should be previously known for which connectives (T, S, N, J) holds, and if they verify the conditions at which the scheme can be submitted, such as it is the nonnull character of the premises' conjunction.

17.2. To answer the question about the validity of the disjunctive scheme,

$$\mu + \sigma, \;\; \sigma' : \mu,$$

it should be analyzed if the functional inequality

$$T(S(a,b),N(b)) \leq a,$$

can be solved. Its solutions, if existing, can actually be used provided the conjunction $T(N(b), S(a, b))$ were not always null.

Notice that for the triplet (min, max, 1 − id), the scheme does not hold, because, for instance, min(max(0.5, 0.3), 1 − 0.3) = 0.5 > 0.3, but it holds with the triplets (W, max, 1 − id), and (W, prod*, 1 − id), because:

- $W(\max(a,b), 1-b) = \max(0, \max(a,b)-1) \leq \max((0, a \cdot b) \leq a,$
- $W(\text{prod} * (a,b), 1-b) = \max(0, a+b-ab-b) = \max(0, a(1-b)) \leq a,$
 with the respective "conjunctions" different from zero.

Hence the disjunctive scheme has no general validity, but depends on the solutions of the former functional inequality. By making $a = 0$, it reduces to the functional equation $T(b, N(b)) = 0$, a well-known one whose solutions are $T = W_f$ and $N \leq N_f$. This is, of course, a necessary but not sufficient condition for the inequality's solutions and, in addition, it is sufficient that $S \leq W_f^*$. With it, and because

$$W_f(S(a,b), N(b)) \leq W_f(W_f^*(a,b), N_f(b)) = f^{-1}(\max(0, \min(1, f(a)+f(b)) - f(b)))$$
$$= f^{-1}(\max(0, \min(1-f(b), f(a)) \leq f^{-1}(\max(0, f(a)) = f^{-1}(f(a)) = a,$$

the former triplets are solutions of the inequality. Hence, although these are not all the inequality's solutions, it is at least solved for the many triplets (T, S, N) verifying $T = W_f$, $S \leq W_f^*$, and $N \leq N_f$, including the two former nondual instances, and also the dual triplet (W, W^*, 1 − id). The disjunctive scheme can be used for reasoning with imprecise words represented by fuzzy sets, at least provided the connectives were on those triplet's family, and it should be remarked that, being $T = W_f$, duality is limited to when it is $S = W_f^*$ and $N = N_f$.

Summing up, if the scheme cannot be freely used in all fuzzy frameworks, there are some algebras where it can be used for formally conducting reasoning with imprecise words.

17.3. What about the scheme of reduction to absurdity,

$$\sigma', \mu' \to \sigma{:}\mu?$$

Now, the corresponding symbolic inequality,

$$\sigma' \cdot (\mu' \to \sigma) \leq \mu,$$

should be posed in terms of the functional inequality,

$$T(N(b), J(N(a), b)) \leq a,$$

for all a, b in [0, 1], and whose left term cannot be zero for all pairs (a, b).

With $a = 0$, it follows that $T(N(b), J(1, b)) = 0$ implying, provided $J(1, b) \neq 0$, $T = W_f$. Namely, if J is such that $J(1, b) = b$ for all b in [0, 1], then and apart from the last specification of T, it suffices that $N \leq N_f$ and $f(J(a, b)) \leq 1 - f(N(b)) + f$

$(N(a))$. Condition $J(1, b) = b$ is verified by many functions J actually used in fuzzy logic.

Note that with $f = \text{id}$, and $N = 1 - \text{id}$, it is $T = W$, and $J(a, b) \leq \min(1, 1 - a + b)$; with $f = \text{id}$, and $N = 1 - \text{id}/1 + \text{id}$, it is also $T = W$, but $J(a, b)$ $(1 - a.(1 - b) + 3b)/(1 + a).(1 + b)$.

Hence the scheme of reduction to absurdity can hold in fuzzy logic in such a way that, although the "and" should be of a very restrictive type, there is a lot of room for both the negation N and the function J representing the conditional.

Summing up, if the scheme is not always valid, there are many algebras where it holds, and in which a formal reasoning using the scheme can be freely used.

Note that for formally using both schemes of disjunction, and reduction to absurdity, at least the conditions shown in Sects. 17.1 and 17.2 jointly hold; that is, a formal fuzzy framework is given by $T = W_f$, $S \leq W_f^*$, $N \leq N_f$, and $f(J) \leq 1 - f(N(b)) + f(N(a))$.

17.4. The constructive dilemma in fuzzy logic is

$$\mu + \sigma, \mu \rightarrow \alpha, \sigma \rightarrow \lambda : \sigma + \lambda.$$

It leads to analyzing the symbolic inequality

$$(\mu + \sigma) \cdot (\mu \rightarrow \alpha) \cdot (\sigma \rightarrow \lambda) \leq \sigma + \lambda,$$

through the complex functional inequality

$$T(S(a, b), J(a, c), J(b, d)) \leq S(b, d),$$

for all a, b, c, d in [0, 1], with the left term different from zero, and the unknowns T, J, and S.

Provided J were fixed by $J(a, b) = S(N(a), b)$, and T were consistent with N, that is, $T(a, b) = 0$ implies $b \leq N(a)$, it is not so difficult to prove that it should be $T = W_f$, $N = N_f$, and $S = \max$. In fact, in the former functional inequality, $a = c = d = 0$ implies $T(b, N(b)) = 0$, and $d = b = 0$ implies $W_f(a, S(N(a), c)) = T(a, c)$, from which when $a > N(c)$ follows $S(N(a), c) = c = \max(N(a), c)$, and therefore $S = \max$. The reciprocal is immediate by just checking the solution.

Hence, the constructive dilemma with $J(a, b) = S(N(a), b)$, holds in fuzzy frameworks given by standard algebras with $T = W_f$, $S = \max$, and $N = N_f$. Only in these frameworks can the constructive dilemma be freely used in a formal deductive reasoning with the imprecise words represented by membership functions.

It should be pointed out where the defined consistency of T with N actually comes from; it is from a typically Boolean property. It is known that in ortholattices contradiction implies incompatibility, but what is not always the case is that $p \cdot q = 0$ implies $p \leq q'$, that incompatibility implies contradiction, except in Boolean algebras where the law of perfect repartition allows it. This is what consistency adds for finding the solutions of the former functional inequality.

Note that, for jointly using the three schemes of disjunction, reduction to absurdity, and the constructive dilemma, the triplet should be (W_f, max, N_f), because max is the minimum t-conorm, and $J(a, b) = \max(1 - a, b)$ actually verifies the condition of Sect. 5.2; that is, $\max(1 - a, b) \leq 1 - (1 - b) + (1 - a) = (1 - a) + b$. Of course, the problem is not yet completely solved by this solution inasmuch as it lacks analyzing what happens with other functions J among those fuzzy logic uses.

Regarding the destructive dilemma,

$$\mu' + \sigma', \alpha \to \mu, \lambda \to \sigma{:}\alpha' + \lambda',$$

the corresponding functional inequality has the same solutions as that for the constructive dilemma, and just provided J were contra symmetrical, that is, $J(a, b) = J(N(b), N(a))$, $J = J \circ (N \times N)$, as it happens in the Boolean case with $p \to q = p' + q$.

17.5. The solutions just reached allow some comments on approximate imprecise reasoning.

(a) With the constructive dilemma's solutions, and provided it were

- $\varepsilon \leq (\mu \to \sigma)(x, y) = \max(1 - \mu(x), \sigma y))$,
- $\delta \leq (\lambda \to \alpha)(x, y) = \max(1 - \lambda(x), \alpha(y))$,
- $\theta \leq (\mu \to \lambda)(x, y) = \max(1 - \mu(x), \lambda(y))$, with ε, δ, θ in [0, 1], it follows that

$$W(\varepsilon, \delta, \theta) \leq \max(\alpha(x), \sigma(y)), \quad \text{or}$$
$$\max(0, \varepsilon + \delta + \theta - 2) \leq \max(\alpha(x), \sigma(y)),$$

which, with the condition $2 < \varepsilon + \delta + \theta$, shows the approximation

$$0 < \varepsilon + \delta + \theta - 2 \leq \max(\sigma(y), \alpha(x)).$$

(b) With the approximate disjunctive scheme,

$$\varepsilon \leq W * (\mu(x), \sigma(y)), \delta \leq 1 - \mu(x),$$

it follows that
$W(\varepsilon, \delta) = \max(0, \varepsilon + \delta - 1) \leq W(W^*(\mu(x), \sigma(y)), 1 - \mu(x)) = \min(1 - \mu(x), \sigma(y)) \leq \sigma(y)$, which, provided $1 < \varepsilon + \delta$, would add to $\mu(x) \leq 1 - \delta$ the bounding $\varepsilon + \delta - 1 \leq \sigma(y)$.

(c) With the approximate scheme of resolution,

$$\varepsilon \leq \max(\mu(x), \sigma(y)), \delta \leq \max(1-\sigma(x), \lambda(y)),$$

follows

$$W(\varepsilon, \delta) = \max(0, \varepsilon + \delta - 1) \leq \max(\mu(x), \lambda(y)).$$

This bounding is significant provided $1 \leq \varepsilon + \delta$, in which case would be $0 < \varepsilon + \delta - 1 \leq \max(\mu(x), \lambda(y))$.

(d) With the approximate reduction to absurdity,

$$\varepsilon \leq 1-\sigma(y), \delta \leq J(1-\mu(x), \sigma(y)),$$

with J such that $J(1, b) = b$, and $J(a, b) \leq 1 - a + b$, it follows that

$$\begin{aligned} W(\varepsilon, \delta) &= \max(0, \varepsilon + \delta - 1) \leq W(1-\sigma(y), J(1-\mu(x), \sigma(y)) \\ &= W(1-\mu(x), \sigma(y) + \mu(x)) = \mu(x), \end{aligned}$$

which, provided $1 \leq \varepsilon + \delta$, would facilitate bounding $0 < \varepsilon + \delta - 1 \leq \mu(x)$. It translates an approximate reasoning such as

If "not p" is up a level $\varepsilon > 0$, "if not q, then p" would be up to a level $\delta > 0$, and if it is $1 \leq \varepsilon + \delta$, q would be up to the level $\varepsilon + \delta - 1 > 0$.

17.6. Let's make a comment on the difference between approximate reasoning with imprecise words, and with precise words. In the second, the design of what intervenes is limited to specifying crisp sets for its precise linguistic terms, and to finding a suitable probability, because usually approximation refers to the uncertainty surrounding the statements or events. For instance, provided it were known that it is $\varepsilon \leq \text{prob}(p)$, and $\delta \leq \text{prob}(p \to q)$, for a continuous t-norm T it would follow that $T(\varepsilon, \delta) \leq T(\text{prob}(p), \text{prob}(p' + q))$, and it should be observed that only with $T = W$ can the second member be easily continued for arriving at prob(q):

$$\begin{aligned} \mathrm{W}(\text{prob}(p), \text{prob}\ (p' + q)) &= \max(0, \text{prob}\ (p) + \text{prob}(p' + q) - 1) \\ &= \max(0, \text{prob}\ (p \cdot (p' + q)) + \text{prob}(p + p' + q) - 1))) \\ &= \max(0, \text{prob}(p \cdot q)) = \text{prob}(p \cdot q) \leq \text{prob}(q). \end{aligned}$$

Hence, $W(\varepsilon, \delta) = \max(0, \varepsilon + \delta - 1) \leq \text{prob}(q)$ is obtained, that, provided it were $1 < \varepsilon + \delta$, would lead to the bounding $0 < \varepsilon + \delta - 1 \leq \text{prob}(p)$, that is, to an approximate probabilistic modus ponens. Note that $\varepsilon = \delta = 1$ leads to $1 = \text{prob}(p)$, which means a probabilistic recovering of the crisp situation.

Analogously, if what is known is $\varepsilon \leq \text{prob}(q')$, and $\delta \leq \text{prob}\ (p' + q)$, it would follow that

$$\varepsilon \cdot \delta \leq \text{prob}(q') \cdot \text{prob}(p' + q) = \text{prob}(q') \cdot (\text{prob}(p') + \text{prob}(q) - \text{prob}(p' \cdot q))$$
$$\leq \text{prob}(p') + \text{prob}(q) - \text{prob}(p' \cdot q) \leq \text{prob}(p'), \text{ since } p' \cdot q$$
$$\leq q \text{ implies } \text{prob}(q) - \text{prob}(p' \cdot q) \leq 0.$$

Hence

$$\varepsilon \cdot \delta \leq \text{prob}(p'),$$

reflects the approximate probabilistic modus tollens. Obviously $\varepsilon = \delta = 1$ gives $1 = \text{prob}(p')$, leading to probabilistically recovering the crisp situation.

Notwithstanding, and as formerly said, the calculus of probability does not interpret prob (If p, then q) by $\text{prob}(p' + q)$, but by $\text{prob}(q/p)$, hence both probabilistic MP and MT deserve to be posed in such form. For instance, for MP, $\varepsilon \leq \text{prob}(p)$, $\delta \leq \text{prob}(q/p)$, leads to

$$\varepsilon \cdot \delta \leq \text{prob}(p \cdot q) \leq \text{prob}(q),$$

which is a better bounding than the former, inasmuch as it is $W(\varepsilon, \delta) \leq \varepsilon.\delta$. In the case of MT, from $\varepsilon \leq \text{prob}(q')$ and $\delta \leq \text{prob}(q/p)$ a bounding for $\text{prob}(p')$ should be found.

As a last probabilistic example, the probabilistic scheme of reduction to absurdity:

$$\varepsilon \leq \text{prob}(q'), \delta \leq \text{prob } (p' \to q) = \text{prob}(p + q),$$

leads to

$$W(\varepsilon, \delta) \leq W(\text{prob } (q'), \text{prob}(p + q)) = \max(0, \text{prob}(q') + \text{prob}(p + q) - 1)$$
$$= \max(0, \text{prob}(p + q) - \text{prob}(q)) = \text{prob}(p + q) - \text{prob}(q)$$
$$= \text{prob}(p) - \text{prob}(p \cdot q) \leq \text{prob}(p).$$

Hence provided $1 \leq \varepsilon + \delta$, it is $0 < \varepsilon + \delta - 1 \leq \text{prob}(p)$. For instance, with $\varepsilon = \delta = 0.9$, it results in $0.8 \leq \text{prob}(p)$, or $\text{prob}(p') \leq 0.2$. Obviously, $\varepsilon = \delta = 1$ gives $1 = \text{prob}(p)$, a probabilistic recovering of the crisp situation.

This scheme also deserves to be posed under conditional probability, that is, with $\varepsilon \leq \text{prob}(q')$, $\delta \leq \text{prob}(q/p')$: $\varepsilon.\delta \leq (\text{prob}(q')/\text{prob}(p')).\text{prob } (q \cdot p')$, from which a bounding should be found for $\text{prob}(p)$.

17.7. A scheme of classical deductive reasoning not considered in the above sections, but still deserving a comment, is the transitive scheme, or "chaining syllogism": $p < q$, $q < r$: $p < r$, of which something was said formerly.

In the Boolean model, this scheme always holds, because $(p' + q) \cdot (q' + r) = p' \cdot q' + p' \cdot r + q \cdot r \leq p' + r$. Concerning the fuzzy model $\mu \to \sigma$, $\sigma \to \lambda$:

$\mu \to \lambda$, already taken into account, it seems of some interest to look at its approximate version, consisting in the following.

Once the boundings are known,

$$(\mu \to \sigma)(x, y) \geq \varepsilon, \text{and } (\sigma \to \lambda)(y, z) \geq \delta,$$

a bound $\alpha = \alpha\ (\varepsilon, \theta) \leq (\mu \to \lambda)(x, z)$ should be found.

These inequalities can be translated into $J\ (a,\ b) \geq \varepsilon$, $J\ (b,\ c) \geq \theta$ => $J\ (a,\ c) \geq \alpha$, for which the T-transitivity of J is sufficient, in which case it follows that

$$T(\varepsilon, \theta) \leq T(J(a, b), J(b, c)) \leq J(a, c), \text{ with } T(\varepsilon, \theta) > 0 \text{ if } \varepsilon > 0, \text{ and } \theta > 0,$$

showing that $J\ (a,\ b) \in (0,\ T\ (\varepsilon,\ \theta)]$.

- If J is W_f-transitive, the bound $W_f(\varepsilon, \theta) = f^{-1}$ (max(0, $f(\varepsilon) + f(\theta) - 1$)) can be null; it happens if and only if $f(\varepsilon) + f(\theta) \leq 1$, equivalent to $\theta \leq f^{-1}(1 - f(\varepsilon)) = N_f(\varepsilon)$; that is, numbers θ and ε are contradictory with respect to the negation N_f.
- In the null case what is obtained is the uninformative conclusion $J(a, c) \in [0, 1]$, informing of nothing; for having $W_f(\varepsilon, \theta) > 0$, it is necessary and sufficient to have $\theta > N_f(\varepsilon)$, that $N_f(\varepsilon)$ acts as a strict hypothesis for θ, and then the conclusion is the truly informative $J(a, c) \in (0, W_f(\varepsilon, \theta)]$.

These links between ε and θ, like other similar ones that formerly appeared, show that many structural questions on the crossed reasonings between precise or imprecise statements in a universe of discourse, and the reasoning's properties on numbers in the real line, still deserve careful scrutiny for what refers to the properties each one can show. For instance, perhaps it could be suitable to start such a scrutiny with probabilistic reasoning, where the strong Boolean laws to which the involved precise statements/events are subjected, and the laws probability enjoys (additivity in particular), can allow a better understanding of the former, perhaps "cryptic," statement.

17.8. Translating the schemes of deductive reasoning into a mathematical representation's framework offers very different possibilities for their validity in formal reasoning, depending on the abundance or scarcity of structural laws in the framework in which the representation is actually made. For instance, as shown, the four modi of classical reasoning are always valid and equivalent in any Boolean setting, but neither in De Morgan or orthomodular settings, nor in a fuzzy one as is the case shown before with MP and MT. At each setting, reasoning cannot be done under the same schemes as in the crisp case, and a suitable framework should be searched for it.

Of course, although the diverse algebras of fuzzy sets follow less defining laws, or axioms, than ortholattices and De Morgan algebras, and, for instance, the four modi either cannot be defined by the proto-form $p \to q = p' + q$, or do not hold, it is proven that for some BAF the schemes can also hold with imprecise statements. It should be recalled that the basic schemes were justified in the general setting introduced in this book, and that it is mainly the lack of <-transitivity that can cause its failing.

What is still to be explored is if, with $p \to q$ different of $p' + q$, those modi can hold. For instance, as was noted, in a De Morgan algebra with $p \to q = p \cdot q$, it is $p \cdot (p \cdot q) = p \cdot q \leq q$, and MP holds, but because the conjunction of the premises is null, $q' \cdot (p \cdot q) = 0$, and although if it is obviously less than or equal to p' ($\leq p'$), MT cannot be freely used. In addition, the scheme of disjunctive reasoning is not yet completely studied when the inclusive or, +, is substituted by the exclusive one, $p \, \Delta \, q = (p + q) \cdot (p \cdot q)'$, as is often done in plain language. For instance, and as can be easily seen with examples, the disjunctive scheme cannot always hold in orthomodular lattices.

The analysis of the validity of some schemes is still an open problem of some interest for doing, computationally, ordinary reasoning with precise and imprecise concepts, and once plain language is represented in fuzzy terms.

For transforming reasoning in a calculus, as Leibniz supposed it can be done, and although restricted to some parts of language, such a calculus requires a specific mathematical framework that, when confronting imprecise concepts represented by the membership functions of the linguistic collectives (the fuzzy sets their linguistic labels generate) requires, as always, previously following a process for well designing the meanings of the involved predicates, connectives, relationships, schemes, and the like. Anyway, a small window for computers is opened by the schemes' general validity with forward, backward, forward–backward, and backward–forward movements.

17.9. A short comment concerning classical crisp transitivity of conditionals is still in order. It was shown that the usual conditional operation $p \to q = p' + q$, is transitive with respect to the Boolean conjunction $(\cdot)$, implying, in the tautological case $p \leq q \Leftrightarrow p' + q = 1$, that the order relation $\leq$ is transitive, because $p \leq q \;\&\; q \leq r \Leftrightarrow p' + q = 1 \;\&\; q' + r = 1 \Rightarrow 1 = (p' + q) \cdot (q' + r) = p' \cdot q' + p' \cdot r + q \cdot \mathrm{r} \leq \mathrm{p}' + \mathrm{r}$, or $1 = p' + r \Leftrightarrow p \leq \mathrm{r}$.

With the conjunctive conditional operation $p \to q = p \cdot q$, if $q \to r = q \cdot r$, it also follows that $(p \cdot q) \cdot (q \cdot r) = p \cdot q \cdot r \leq p \cdot r = p \to r$, showing that it is also transitive, and that in the tautological case $p \to q = 1 \Leftrightarrow p = q = 1$, the associated binary relation is also transitive. Analogously, with the conditional operation $p \to q = p \cdot q + p' \cdot q'$, $q \to r = q \cdot r + q' \cdot r'$, follows

$$(p \cdot q + p' \cdot q') \cdot (q \cdot r + q' \cdot r') = p \cdot q \cdot r + p' \cdot q' r' \leq p \cdot r + p' \cdot r' = p \to r.$$

In the same vein, with $p \to q = q$, and $q \to r = r$, it follows that $q \cdot r \leq r = p \to r$.

Notwithstanding, the question whether all the Boolean conditionals are transitive has a negative answer. For instance, with the conditional operation

$$p \rightarrow q = p \cdot q, \quad \text{if } p \cdot q \neq 0,$$

and

$$p \rightarrow q = p' + q, \quad \text{if } p \cdot q = 0,$$

it is $(p \rightarrow q) \cdot (q \rightarrow r) = q \cdot r$ (provided $p \cdot q = 0$ and $q \cdot r \neq 0$), but it is not $q \cdot r \leq p \rightarrow r$ in all cases, as it is clear if $p \cdot r \neq 0$ because it is not necessarily $q \cdot r \leq p \cdot r$. Hence, in the classical case, not all conditional operations show transitivity; this scheme does not universally hold for conditionals.

17.10. Not all the classical schemes were considered in Chaps. 4 and 5. For instance, nothing has been said regarding the schemes of

- Importation, $p \rightarrow (q \rightarrow r) = p \cdot q \rightarrow r$,

and

- Exchange, $p \rightarrow (q \rightarrow r) = q \rightarrow (p \rightarrow r)$,

whose validity and equivalence in the classical case is proven by

- With $p \rightarrow q = p' + r$: $p \rightarrow (q \rightarrow r) = p' + (q' + r) = (p' + q') + r = (p \cdot q)' + r = p \cdot q \rightarrow r$, and $p' + (q' + r) = q' + (p' + r) = q \rightarrow (p \rightarrow r)$,
- With $p \rightarrow q = p \cdot q$: $p \rightarrow (q \rightarrow r) = p \cdot q \rightarrow r = p \cdot q \cdot r$.

Their validity in fuzzy logic depends, obviously, on the particular BAF in which function J and conjunction T are used. For instance,

- With $T = \min$, and $J(a, b) = \max(1 - a, b)$, it is

$J(\min(a, b), c) = \max(\max(1 - a, 1 - b), c) = \max(1 - a, 1 - b, c)$, and $J(a, J(b, c)) = \max(1 - a, \max(1 - b, c)) = \max(1.a, 1 - b, c)$, and the classical scheme holds.

- With $T = \text{prod}$, and the same J, it is

$J(a.\, b, c) = \max(1 - a.\, b, c)$, and $J(a, J(b, c)) = \max(1 - a, 1 - b, c)$, that are not coincidental and the scheme fails. For instance, $c = 0$ shows the first equal to $1 - a.\, b$, but the second equal to $\max(1 - a, 1 - b)$, and it suffices $a = b = 1/2$, to have the first equal to 3/4, and the second equal to 1/2.

Nevertheless, if these schemes are well and deeply analyzed in the framework of the current theoretical fuzzy logic, the proving of their general validity with the primitive relation $<$ of natural inference is only immediate for the scheme of importation $[p < (q < r)] \approx [p \cdot q < r]$ under transitivity. In fact, provided $p \cdot q < q$, with $q < r$ being attainable from p, and presuming the triplet $(p \cdot q, q, r)$ is transitive, it would be $p \cdot q < r$.

Summing up, because the classical schemes of crisp reasoning are not universally valid with imprecise words in a fixed framework, they should be carefully used in the mathematical frames where ordinary reasoning can be represented.

That many schemes hold in the general setting of natural inference does not mean that in a formal representation frame with fuzzy sets they all should hold, but that for each of them there should exist some BAFs where it holds. What can be difficult is to find a particular functionally expressible BAF for a given scheme.

References

1. E. Trillas, C. Alsina, E. Renedo, On some schemes of reasoning in fuzzy logic. New Math Nat Comput **7**(3), 433–451 (2011)
2. M. Baczynski, B. Jarayam, *Fuzzy implications* (Springer, Berlin, 2008)
3. D Dubois, R Martin-Clouaire, H Prade, *Practical Computations in Fuzzy Logic*. ed. by M.M. Gupta et al. Fuzzy Computing (Elsevier, Amsterdam, 1988), pp. 11–34
4. E. Trillas, L. Eciolaza, *Fuzzy logic* (Springer, Berlin, 2015)
5. I. García-Honrado, E. Trillas, Remarks on the symmetrical difference from an inferential point of view. Multiple-Valued Logic Soft Comput **24**(1-4), 35–59 (2015)

Chapter 18
Questions on Monotony

The existing or lacking character of monotony is, to some extent, an important characteristic of reasoning. In general, monotony concerns how conclusions vary when evidence increases, when more "safe" premises are added. When the number of conclusions cannot decrease, it is said that reasoning shows monotony, when they cannot increase shows antimonotony, and when it cannot be foreseen, that is, when there is no law in such respect, shows nonmonotony.

Indeed, information actually flows and hence reasoning is not a static, but a dynamic, process, often supported on some relationships between the premises and the conclusions that can be expected. Because reasoning is mainly done under conditionality, analyzing either the monotonic or nonmonotonic character of conditionals has some interest.

This chapter is devoted to reflections on this topic for both precise and imprecise concepts represented, respectively, in a Boolean framework, and in a standard BAF (basic fuzzy algebra). A study, currently incomplete, on monotonic relations and, particularly, on monotonic conditionals whose managing is basic for reasoning, could help us better understand in what nonmonotonic reasoning actually consists. Provided monotonic conditionals were characterized, it would lead to determining when a conditional is nonmonotonic; even incomplete, such is the goal of this chapter.

18.1. Let's consider a crisp conditional relation in a set X, that is, a binary relation $< \subseteq X \times X$, submitted to verify the modus ponens (MP) scheme: $p, p < q$: q, but seeing it from the point of view of the existence of subsets V of X, verifying:

$$\text{If } p \in V \text{ and } (p, q) \in <, \quad \text{or} \quad p < q, \text{ then } q \in V,$$

as a way of expressing modus ponens. The family of these subsets V is not empty; at least, one of them is, obviously, X; they just generalize the set of true elements in the classical calculus, and the capital letter V stands for "veritable".

E. Trillas, *On the Logos: A Naïve View on Ordinary Reasoning and Fuzzy Logic*,
Studies in Fuzziness and Soft Computing 354, DOI 10.1007/978-3-319-56053-3_18

The binary relation $<$ is said to be $\cdot$—monotonic provided,

- X is endowed with an operation $(\cdot)$ representing the linguistic "and",
- If $p < q$, then $p \cdot r < q$, for all r in X,

that is, the new information furnished by r does not change the consequent q after the conjunction of the antecedent p with r. Once any r is "conjunctively added" $(p \cdot r)$ to the former antecedent, the conjunction keeps q as a consequent under $<$, although some new consequent can also appear.

The sets V of veritable elements can be called "states" of the conditional; those sets are preserving "the veritable". For instance, if X were an ortholattice or a De Morgan algebra with $<$ equal to its lattice's order $\leq$, one of its states would be $V = \{1\}$, only containing the lattice's maximum, representing the element that allows propagating total truth under the relation $< = \leq$; in the case of a Boolean algebra, and also with $< = \leq$, the natural lattice's order, and for $V = \{1\}$, such elements are the tautologies. Obviously, if $p = 1$ and $p \leq q$ $(1 \leq q)$, then $q = 1$.

Note that in lattices, and with $< = \leq$, it is always $p \leq p$, for all p, and it is "$p \leq q \Rightarrow p \cdot r \leq q$", because $p \cdot r \leq p$, and $\leq$ is transitive; the natural partial order of lattices is $\cdot$—monotonic. It represents a particular instance because, as commented, $<$ is not always transitive in plain language although it is taken as reflexive, but it also holds in general provided it holds the property $p \cdot r < p$, because it is always accepted that $p < p$. Under this supposition the natural inference relation $<$ is $\cdot$—monotonic.

Note that taking the bivalued membership, or characteristic, function $\mu_<$: $X \times X \to \{0, 1\}$ of $<$, instead of it, $\cdot$—monotony can be defined in the equivalent and compact form,

$$\mu_<(p, q) \leq \mu_<(p \cdot r, q), \text{ for all } p, q, r \text{ in } X.$$

Then, the states V are those whose bivaluate membership function μ_V is such that if $\mu_<(p, q) = 1$, and $\mu_V(p) = 1$, then it is $\mu_V(q) = 1$ that, in its turn, can be compacted in the single and equivalent inequality,

$$\mu_<(p, q) \leq \min(\mu_V(p), \mu_V(q)), \text{ for all } p, q \text{ in } X,$$

showing that V is a state for $<$ if and only if the characteristic function of such relation is upper bounded by the Cartesian product relation $V \times V$, defined by $(\mu_V \times \mu_V)(p, q) = \min(\mu_V(p), \mu_V(q))$, that is, if it is $< \subseteq V \times V$. Provided it were $\mu_<(p, q) = 1$, it would imply $1 = \mu_V(p) = \mu_V(q)$, or $p \in V$, and $q \in V$, for all p, $q \in V$. Provided it were $\mu_<(p, q) = 0$, then p, and q, would be free of being or not in V.

Provided $p \cdot r \in V$ were always to imply $p \in V$, and $r \in V$, then the relation $R_V = (V \times V) \cup (V^c \times X)$, would be $\cdot$—monotonic. Hence, it suffices a subset W of X such that there is an instance $p \cdot r \in W$ such that either p is not in W, or r is not in W, to have the nonmonotonic relation $R_W = (W \times W) \cup (W^c \times X)$. Note that it is

$$(p, q) \in R_W \Leftrightarrow (p \in W \text{ and } q \in W), \text{ or } (p \text{ is not in } W),$$

mimicking the expression $p' + p \cdot q$ that, as is well known, is equivalent in Boolean algebras to $p' + q$.

18.2. What about probabilistic relations in a Boolean algebra?

For instance, is it that conditional probability establishes a monotonic relation? Is it prob $(p/q) \leq$ prob$(p \cdot r/q)$ for all r in the Boolean algebra? Conditional probability can be seen as reflexive, because prob $(p/p) = 1$, but the answer is, obviously, not, and to see it suffices to take r such that $r \leq p' \Leftrightarrow r \cdot p = 0$, because then the second probability is zero, but the first is not necessarily null. Thus, conditional probability is not $\cdot$—monotonic.

Another nonmonotonic example, but with a precise relation in a Boolean algebra endowed with a probability prob, is the following.

$$p < q \Leftrightarrow \text{Either prob } (p) = 0, \text{ or prob } (p) > 0, \text{ and prob } (q/p) > 0,$$

with which,

- If prob $(p) = 0$, it would be $p < x$ for all x in the Boolean algebra, and, in particular, $p < p$.
- If prob $(p) > 0$, it would be prob (p/p) = prob$(p \cdot p)$/prob (p) = prob (p)/prob $(p) = 1 > 0$, and $p < p$.

Consequently, it is always $p < p$, and $<$ is a reflexive relation, but $<$ is not $\cdot$—monotonic. Indeed, if $p < q$, it suffices to choose an element r such that prob $(p \cdot r) > 0$, and prob $(p \cdot q \cdot r) = 0$, with which it is prob $(q/p \cdot r) = 0$, and it is not $p \cdot r < q$.

18.3. What about with respect to the states for crisp conditionals, that is, the sets V allowing propagation of the veritable character by modus ponens?

It depends, of course, on the conditional definition; for instance, in an ortholattice with $< = \leq$, all order "intervals" $[a, 1] = \{x;\ a \leq x \leq 1\}$, are states, because it is $a \leq b$ if and only if $b \in [a, 1]$. Notice that the union of all these order intervals is X; in fact, it is obvious that such a union is contained in X, and if $a \in X$, $[a, 1]$ were contained in X, and $[0, 1]$ were also contained in such a union. Their intersection is obviously $\{1\}$.

In the case of a standard algebra, a fuzzy relation R: $[0, 1]^X \times [0, 1]^X \to [0, 1]$, with a continuous t-norm T representing the conjunction "and", can be defined to be T-monotonic, provided

$$R(\mu, \sigma) \leq R(T(\mu, \lambda), \sigma), \text{ for all } \mu, \sigma, \lambda \in [0, 1]^X,$$

and its fuzzy states as those $\alpha \in [0, 1]^X$, such that, and analogously, the relation $\alpha \times \alpha$ is an upper bound of R, that is,

$$R(\mu, \sigma)(x, y) \leq \min(\alpha(x), \alpha(y)), \text{ for all } x, y \text{ in } X.$$

Note that the numerical values of $R(\mu, \sigma)$ reflect how strong, or weak, the link is between μ and σ at each point (x, y).

Instead of relations such as R, the fuzzy case often deals with "operations", such as those coming from expressing conditionals by statements; for instance, when it is understood "If μ, then σ" ($\mu \to \sigma$) by "not μ, or σ" ($\mu' + \sigma$), or as "μ and σ" ($\mu \cdot \sigma$). In these cases, the operation $\to$ is $\cdot$—monotonic provided it were

$$\mu \to \sigma \leq \mu \cdot \lambda \to \sigma, \text{ for all } \mu, \sigma, \lambda \text{ in } [0, 1]^X,$$

and its states can be defined as those fuzzy sets α, such that

$$(\mu \to \sigma)(x, y) \leq \min(\alpha(x), \alpha(y)).$$

For instance, the definition $\mu \to \alpha = \mu \cdot \sigma$, giving a conjunctive conditional operator with $T = \min$, $\min(\mu, \mu \cdot \sigma) \leq \mu \cdot \sigma \leq \sigma$, is not min-monotonic because it is $\mu \cdot \sigma \cdot \lambda \leq \mu \cdot \sigma$. Inasmuch as the same inequality holds for some operations $* \leq \min$, as are all t-norms, the conjunctive conditional is never $*$—monotonic for those operations. Hence, working with a conjunctive conditional, like working with the conditional probability, could easily lead to nonmonotony, and to not deducing. What follows from $\mu \cdot \sigma$ also follows from $\mu \cdot \sigma \cdot \lambda$ that is less than or equal to $\mu \cdot \sigma$, but there can exist fuzzy sets following this second and not the first; after conjunction with λ, nonpreviously existing consequences can appear.

What about those fuzzy conditionals $\mu \to \sigma$ that can be represented by operators J of the types $S(N(\mu), \sigma)$, and $J_T(\mu, \sigma)$?

The second are reflexive, because it is $J_T(a, b) = 1 \Leftrightarrow a \leq b$, a property not fulfilled by the first; for instance, with the operator max $(1 - a, b)$, it is $\mu_{1/4} \to {}_{1/4} = \mu_{\max(1 - 1/4,\, 1/4)} = \mu_{3/4}$.

Notwithstanding, all these conditionals are T-monotonic for any continuous t-norm T, because they verify the decreasing property for antecedents:

$$a \leq b \Rightarrow J(a, c) \geq J(b, c) \text{ for all } c \text{ in } [0, 1],$$

and hence because $\mu \cdot \lambda \leq \mu$, it is $J(\mu \cdot \lambda, \sigma) = \mu \cdot \lambda \to \sigma \geq \mu \to \sigma = J(\mu, \sigma)$, with $\alpha \cdot \beta = T(\alpha, \beta)$ for all α, β in $[0, 1]^X$, and any t-norm T. For instance,

- If $a \leq b$, because $N(b) \leq N(a)$, it follows that $S(N(b), c) \leq S(a, c)$, or $b \to c \leq a \to c$.
- Because $T(z, a) \leq T(z, b)$, it is $\{z \in [0, 1]; T(z, b) \leq c\} \subseteq \{z \in [0, 1]; T(z, a) \leq c\}$, and $J_T(a, c) \geq J_T(b, c)$.

All these conditionals allow dealing with $\cdot$—monotony.

18.4. What about with respect to the states for both the S and the conditional operators J_T?

For instance, in the case of an S operator, its states are those α such that $S(N(\mu(x)), \sigma(y)) \leq \min(\sigma(x), \alpha(y))$, from what follows $N(\mu(x)) \leq \alpha(x)$, and $\sigma(y) \leq (y)$, and it suffices to find those α verifying $N(\alpha(x)) \leq \mu(x)$, and $\sigma(x) \leq \alpha(x)$, or $N(\alpha(x)) \leq N(\sigma(x))$, that is, $\alpha' \leq \mu$ and $\alpha' \leq \sigma'$, or $\alpha' \leq \min(\mu, \sigma') \Leftrightarrow \max(\mu', \sigma) \leq \alpha$. The states are among the fuzzy sets in the pointwise order interval $[\max(\mu', \sigma), \mu_1]$.

Regarding J_T operators, the states α should verify $J_T(\mu(x), \sigma(y)) \leq \min(\alpha(x), \alpha(y))$. Thus,

- With $T = W$, it is $J_W(\mu, \sigma)(x, y) = \min(1, 1 - \mu(x) + \sigma(y)) \leq \min(\alpha(x), \alpha(y))$, from where it would follow that $1 = \min(\alpha(x), \alpha(y))$, if $\mu(x) \leq \sigma(y)$, and $1 - \mu(x) + \sigma(y) \leq \min(\alpha(x), \alpha(y))$, if $\mu(x) > \sigma(y)$. Hence, the states are among those α such that for some y is $\alpha(y) = 1$, and $1 - \mu(x) + \sigma(x) \leq \alpha(x)$ for the other x.
- With T = prod, it is $J_{prod}(a, b) = \min(1, a/b)$, and $(\mu \to \sigma)(x, y) = \min(1, \mu(x)/\sigma(y)) \leq \min(\alpha(x), \alpha(y))$. Hence, if $\mu(x) \leq \sigma(y)$ is $1 = \min(\alpha(x), \alpha(y))$ and if $\mu(x) > \sigma(y)$ it would be $\mu(x)/\alpha(y) \leq \min(\alpha(x), \alpha(y))$, with which if there is an x such that $\alpha(x) = 1$, and the states are among those α verifying $\mu(x)/\sigma(x) \leq \alpha(x)$, that is, those $\alpha \in [0, 1]^X$, whose values belong to the closed numeric interval $[\min(1, \mu(x)/\sigma(x)), 1]$.
- With $T = \min$, because it is $J_{\min}(a, b) = 1$, if $a \leq b$, and $J_{\min}(a, b) = b$, if $a > b$, it would follow that $1 = \min(\mu(x), \sigma(y))$ if $\mu(x) \leq \sigma(y)$, and $\sigma(y) \leq \min(\alpha(x), \alpha(y))$ if $\mu(x) > \sigma(y)$. That is, the states should be searched among the $\alpha \in [0, 1]^X$ such that $\alpha \leq \sigma$, among the α in the interval $[\mu_0, \sigma]$.

18.5. Even if what is in this section were not only incomplete, but also insufficient, it seems that the study of either nonmonotonic relations, or operations, deserves to be continued, mainly, for counting with the possibility of recognizing which type of nonmonotonic relation, or operation, can be suitable for representing a linguistic relationship whose praxis clearly appears as a nonmonotonic one. In fact, such is an open subject currently only closed for preorders and fuzzy preorders.

References

1. A. Sobrino, E. Trillas, Can fuzzy logic help to pose some problems in the philosophy of science?, in *Representations of Scientific Rationality*, ed. by A. Ibarra et al. (Rodopi, Amsterdam, 1997), pp. 277–300
2. E. Trillas, S. Cubillo, On monotonic fuzzy conditionals. J. Appl. Non-Class. Logics **4**(2), 201–214 (1994)
3. E. Trillas, On membership and fuzzy logic states, in *Proc. 2nd European Congress on Intelligent Techniques and Soft Computing* (1994), pp. 796–801
4. S. Cubillo, E. Trillas, Characterizing non-monotonic fuzzy relations. Soft Comput. **1**(4), 162–165 (1997)
5. S. Cubillo, On some inexact relations in probabilized Boolean algebras. Mathware Soft Comput. **5**, 39–47 (1998)

References

Chapter 19
Questions on "Not Covered By"

When fuzzy sets were introduced in 1965, the only back referents to endow them with some theory were those theories on classical sets, on classical predicate logic, and those on the multiple-valued logics. These last were used for the design of electrical circuits not only closing or opening its gates.

Instead, more than 50 years later, when the challenge consists in developing a theory on which can be based Zadeh's idea on computing with words (CwW), what should be taken into account as referents are the many aspects and problems of plain language and ordinary reasoning not well covered by sets and formal deduction. New points of view seem to be necessary to face CwW's theoretical future and mainly for foreseeing which kind of research, and on which topics, it should be conducted for grounding CwW in an experimental science of language and reasoning, that is, in a kind of physics of language and reasoning. One such topic, for instance, is that concerning what can be understood in plain language by "not totally covered by P", or "partially covered by not P", and how they can be represented for what refer the elements in the universe of discourse and further than membership functions.

This is a question that, with a long history, could be thought about in several forms but whose first and basic problem is with Q naming "not totally covered by P" in a universe X, to capture those relations $<_Q$ that eventually can coincide with $<_{P'}$ or with $<_P{}^a$, when Q can be identified with either not P, or with an opposite of P, or with another word used in language. A linguistic example of such words is done by the predicate P^m = not P and not $P^a = P' \cdot (P^a)'$, usually known as the "middle term" of P, and once accepting that the negation of the antonym has sense: for example, medium for "big and not short", or warm for "hot and not cold". The problem partially lies in founding "names", existing in plain language, for such predicates Q, such as the former "medium" and "warm", on analyzing their uses or meanings and particularly those of $(P^a)'$ and the medium term, as well as on introducing mathematical models testable against the reality of language, and consequently accepted or refused in each contextual situation. Although all this is, indeed, a currently open subject, a reflection on it is in order.

E. Trillas, *On the Logos: A Naïve View on Ordinary Reasoning and Fuzzy Logic*,
Studies in Fuzziness and Soft Computing 354, DOI 10.1007/978-3-319-56053-3_19

19.1. Remember that it is $<_{P^a} = <_P^{-1} \subseteq <_{P'}$, $<_{P^m} = <_{P' \cdot (P^a)'} = <_{P'} \cap <_{(P^a)'}$, and $(P^a)^a = P$, but that P is not always comparable with $(P')'$, even if sometimes it were to hold "If x is $(P')'$ then x is P", or "If x is P, then x is $(P')'$", and, as a particular situation, P and $(P')'$ can coincide. That is, for instance, at least Q can be understood as covering "not all P", or covering "not totally P", or "partially P". It can be supposed that $<_Q \subseteq <_{P^a} \subseteq <_{P'}$, or that $<_{P^a} \subseteq <_Q \subseteq <_{P'}$, and, consequently, some meaning's measures for Q studied provided, in the first case, it were $<_Q \neq \emptyset$.

The relation of coherence between negation and antonym, $\mu_{P^a} \leq \mu_{P'}$, should hold for all the opposites of P. Thus, provided P' were functionally expressible by a strong negation N_P, the function

$$\text{Sup}\{\mu_{P^a}; \text{for all } P^a\} = K(P) \leq \mu_{P'} = N_P \circ \mu_P$$

can be seen as the "fuzzy kernel of negation", that can, or cannot, coincide with $\mu_{P'}$.

For instance, in $X = [0, 1]$ with P = big, and accepting identification of the symmetries s_{big} of $[0, 1]$ with the strong negations $N_f = f^{-1} \circ (1 - f)$, whose supremum is the (discontinuous) greatest negation function,

$$N_{\min}(x) = 1, \text{if } x \in [0, 1), \text{and } N_{\min}(1) = 0,$$

it is $K(\text{big}) = N_{\min}$, specifying the crisp subset $[0, 1)$. Hence, under these suppositions, the only element in the unit interval that can always be qualified as big is 1, the always accepted prototype of big in $[0, 1]$.

Of course, on the idea of the fuzzy kernel of negation it is neither known for what it can practically serve, nor is it well studied. It is not known if function $K(P)$ could represent an opposite, a negation, or simply one of the above predicates Q. For instance, the function $K(\text{big}) = \mu_{[0, 1)}$ could be seen as representing the linguistic label "totally not big", whose negation $1 - K(\text{big}) = \mu_{\{1\}}$, represents "totally big". Anyway, and for doing such an investigation, more examples in universes different from $[0, 1]$ should be studied.

19.2. The crisp set

$$A(P) = \{x; \mu_P(x) \leq \mu_{P^a}(x)\},$$

containing those elements of X showing P less than they show P^a, that is, those that are less covered by P than by P^a, can be called the "crisp kernel of opposition", and, for instance, if $X = [a, b]$ were a closed interval of the real line, the number Sup $A(P)$ would exist that can be seen as separating what is P from what is P^a.

For instance, with $X = [0, 10]$, P = big, and P^a = small, with $\mu_{\text{big}}(x) = x/10$, and the symmetry $s(x) = 10 - x$, it is

$$x/10 \leq (10-x)/10 \Leftrightarrow x \leq 5;$$

hence A(big) = [0, 5], and Sup A(big) = 5. In [0, 5] there are the elements in [0, 10] that are "less big than small", and they are separated by 5 from those that are "less small than big". Nevertheless, with the different symmetry $s(x) = 10(10 - x/10 + x)$, from $x/10 \leq 10 - x/10 + x \Leftrightarrow x^2 + 20x - 100 \leq 0$, follows A(big) = $[0, 10\sqrt{2} - 1, 10]$, with Sup A(big) = $10\sqrt{2} - 1 \approx 4.1421$. As always, everything depends on the designed terms.

As is obvious, provided P were a precise word in X, and hence specified by a crisp subset, it would not be sure that $A(P)$ could always coincide with the complement of such a subset; it would depends on the symmetry $s_P = s$ taken for obtaining $\mu_P{}^a = \mu_P \circ s$. A question remaining open, is if the subset $A(P)$ of X can substitute, in some parts of ordinary reasoning, the typical complement for precise words, or the negation membership function for the imprecise ones, or can it facilitate a crisp representation of what is not under P^a.

19.3. From the self-contradiction inequality $\mu \leq \mu'$, a crisp kernel of negation can be defined by

$$N(P) = \{x; \mu_P(x) \leq N_f(\mu_P(x))\} = \{x; \mu_P(x) \leq f^{-1}(1/2)\},$$

containing those points in X showing P less than not P, those that are less covered by P than by not P, and the points that are more not P than P. Were $X = [a, b]$, then the number Sup $N(P) \in [a, b]$ would exist, separating what is P from what is not P, and allowing us to see the interval $[a, \text{Sup } N(P)]$ as the crisp kernel of negation.

For instance, if $X = [0, 10]$, P = big, and $\mu_{\text{big}}(x) = x/10$, with the negation $(1 - x)/(1 + x)$ whose fixpoint is $\sqrt{2} - 1$, is $N(\text{big}) = \{x \in [0, 10]; x/10 \leq \sqrt{2} - 1\}$; that is, Sup $N(P) = 10(\sqrt{2} - 1)$, and $N(\text{big}) = [0, 10(\sqrt{2} - 1)]$. With it, the numbers that can be considered as actually big are those that are greater than $10(\sqrt{2} - 1) \approx 4.142$. Note that with the negation $N(x) = 1 - x$, is $x/10 \leq 1 - x/10 \Leftrightarrow x \leq 5$, and $N(\text{big}) = [0, 5]$.

It should be noted that, provided P were precise, the crisp kernel of negation would coincide with its crisp complement. Actually, because in this case $\mu_P(x) \in \{0, 1\}$, and $\mu_{P'}(x) = 1 - \mu_P(x)$ has fixpoint $\mu_{1/2}$, it is $N(P) = \{x; \mu_P(x) \leq ½\} = \{x; \mu_P(x) = 0\}$, the crisp complement of the subset $\{x; \mu_P(x) = 1\}$ specified by P. Hence, if P is imprecise, the set $N(P)^c$, the complement of $N(P)$, can be seen as that containing the elements that are more P than not P.

Note also that provided antonym and negation are coherent, $\mu_P{}^a \leq \mu_{P'}$, it is $A(P) \subseteq N(P)$. It happens, for instance, in the former example with P = big, $N(x) = 1 - x/1 + x$, $s(x) = 10(1 - 10x/1 + 10x)$, and $\mu_{\text{big}}(x) = x/10$. Because it is $N(P)^c \subseteq A(P)^c$, it supports seeing the elements in $N(P)^c$ as those that are properly P, and that are separated from the others by the point Sup $N(P)$.

19.4. There are some theoretically driven fuzzy researchers who derive the negation function N from the conditional functions J_T, with T a continuous t-norm, and by defining $N(a) = J_T(a, 0)$, that reproduces the classical equivalence $p' = p' + 0 = p \to 0$. For continuous t-norms different from an ordinal sum, it is:

- If $T = \min$, or $T = \text{prod}_f$, it would be $J_T(a, 0) = N_{\min}(a)$
- If $T = W_f$, it would be $J_T(a, 0) = f^{-1}(1 - f(a)) = N_f(a)$.

Hence, under this interpretation of the negation, coming from understanding "if p, then q" in a form generalizing "not p or q", strong negations only appear linked to the t-norms in the family of W, that of Lukasiewicz, that is, from a restrictive form of representing conditionals that is reminiscent of classical logic.

Note that representing conditionals by the functions $J(a, b) = T(a, b)$ as is usual in fuzzy control, $J(a, 0) = 0$ for all a in $[0, 1]$ gives no negation function.

The reason lies in that the interests of logic are mainly related to formal languages, but not with the plain ones where no universal form of representing conditional statements exists, and where sometimes it is very difficult to describe perfectly by words the antecedent's negation.

19.5. In this chapter it is only tried to reflect on a topic that fuzzy logic's praxis perhaps manages in a too simplistic form concerning plain language, and in the way towards counting with the calculus necessary for actually arriving at CwW. It refers to a view on what is not properly under a linguistic label, and for what language not only counts with negation, but with opposites, and middle terms. All this served Zadeh for introducing linguistic variables as a form of dealing with something analogous to the crisp partitions, and that in fact generalizes the concepts of coverage and partition to the fuzzy world.

What is exposed in Sects. 19.2 and 19.3 is just for helping theoreticians not only to go towards the true basis on which CwW is grounded, but also to anchoring it in plain language and ordinary reasoning concerning the elements in the universe of discourse, and instead of membership functions. CwW cannot be an isolated mathematical subject, but a broader subject dealing with linguistic imprecision and the nonrandom uncertainty of imprecise statements.

As is apparent, both what is under "crisply covered by", and "crisply not covered by", still deserves more study, because it is neither closed by strong negations nor by symmetries for representing opposites. With the advancing of CwW, the time of considering larger statements than those considered in the current applications sooner or later will arrive and will call for a deeper knowledge of the linguistic separation between what is under a predicative imprecise word, and what is not, that is, calling for a new view of the idea of "linguistic complement", of which this chapter is but a tentative attempt at offering a new and different focus than that currently offered.

References

1. E. Trillas, R. Seising, Turning around the ideas of 'meaning' and 'complement', in *Fuzzy Technology*, ed. by M. Cullan et al. (Springer, Switzerland, 2016), pp. 3–31
2. E. Trillas, A short dialogue concerning 'what is' and 'what is not' with imprecise words, in *Fuzzy Logic and Information Fusion*, ed. by T. Calvo et al. (Springer, Berlin, 2016), pp. 237–242
3. E. Trillas, A.R. de Soto, A reflection on fuzzy complements, Proc. IFSA-EUSFLAT, 823–827 (2015)
4. K. Green, *Bertrand Russell, Language and Linguistic Theory* (Continuum Pubs., Easton, 2007)

Chapter 20
Questions on Sorites in Ordinary Reasoning

The sorites paradox, that seems to have been introduced by Eubulides of Miletus in the fourth century BC, comes from the Greek word, *soros*, heap, and refers to something like the following. If in a heap of wheat there are one million grains and one of them is removed, it still remains a heap, but, removing one grain after the other a moment will arrive at which the heap will disappear, and it will remain only a simple mass of wheat, no longer being a heap. Posed in another way, if a grain is not a heap, and two grains is not a heap, and three grains is not a heap, … , how many grains will actually constitute a heap? Are there a number of grains not constituting a heap, but such that by adding one more grain a heap appears?

For a layperson, viewing a heap only depending on the number of grains but not, for instance, on its three-dimensional shape, can appear as something surprising, inasmuch as it is shape that allows people to recognize what a heap actually is; hence it seems that a reflection on the subject from a point of view closer to that of laypeople can be suitable. In the end how people perceive things is relevant for both plain language and ordinary reasoning; because of it, and instead of only talking about computing with words (CwW), Lotfi A. Zadeh usually and rightly talks about computing with words and perceptions.

20.1. The word heap is used as a vague one, of which it seems difficult to measure its meaning for the obvious difficulty of recognizing if this is "less a heap" than that.

The word "heap" is well anchored in language because people recognize, even with the presence of borderline cases, what a heap is, or is not, and its philosophical analysis was based on a type of argument known as "little by little". But there is a new possibility for seeing it from the point of view of the former "crisp kernels" of negation and opposition. It is also a possibility for trying to look again at the "point of separation" between, for instance, heap and mass; a point separating what is from what is not, as was done in the former chapter, and about which, for instance, the philosopher Max Black stated that it should exist but, simultaneously, that it is impossible to be found. Keeping many doubts about if what exists can be never

E. Trillas, *On the Logos: A Naïve View on Ordinary Reasoning and Fuzzy Logic*, Studies in Fuzziness and Soft Computing 354, DOI 10.1007/978-3-319-56053-3_20

found, a tentative move towards its clarification could be interesting for any scientifically motivated spirit.

Note that the word heap has no clearly accepted antonyms in plain languages such as English and Spanish; neither, for instance, does the word "unheap", nor anything similar exist in Spanish. Nevertheless, and as shown, this does not mean that once a membership function for "heap" was designed, no membership functions for its antonyms can be obtained.

20.2. For trying to see if Black's separation point can be found, let's start with some comments on such a question with the help of the toy example given by the word "small" when playing with numbers in the interval [0, 10]. Let's remember that in plain language's use of "small" it is not enough to state that if x is qualified as small, also are such all $y \leq x$, but that all those z that are very close and greater than x, are also small; provided the number 2.6 can be qualified as small, not only 2.5999 but also 2.6001 could be qualified as small. That is, a number $\varepsilon > 0$ should exist such that if x is qualified as small, all the numbers $z \in (x - \varepsilon, x + \varepsilon)$, are also small; this means that small is flexible, and its quantitative meaning is captured by any continuous, strictly decreasing, membership function μ_{small}.

For instance, if $\mu_{small}(x) = 1 - x/10$, and $x - \varepsilon < z < x + \varepsilon$, it would be $\mu_{small}(x + \varepsilon) \leq \mu_{small}(z) \leq \mu_{small}(x - \varepsilon)$, and it could not be $\mu_{small}(x - \varepsilon) = 0$ because it would imply $\mu_{small}(z) = 0$ for all $z \in (x - \varepsilon, x + \varepsilon)$; that is, it follows the absurd that μ_{small} is not strictly decreasing in the interval $(x - \varepsilon, x + \varepsilon)$. There cannot be points different from $z = 10$ with null value of the membership function μ_{small}, it cannot show jumps, and no characteristic function of a crisp set can represent small.

In addition, that such a predicate cannot truly specify a crisp subset of [0, 10] can be proven as follows.

A set S containing those numbers qualifiable as small actually exists, because 0 is small, and consequently all x in $[0, \varepsilon)$ should also be small; it is $S \neq \emptyset$. Because [0, 10] is a compact set in the topology of the real line, and 10 is not in S, there exists $s = \text{Sup } S$ in [0, 10] such that all the $x < s$ are small, and, hence also s is small, and also those in $[s, s + \varepsilon]$ are small, contradicting that s is the supremum among the small numbers. In principle it was presumed that it is $s \neq 10$, therefore the only possible conclusion is that there is not a crisp subset S containing all numbers in [0, 10] that are qualifiable by small in plain language.

This is what the old sorites paradox shows, and that right now just instantiates that there are words whose use cannot be specified by a crisp set in the corresponding universe of discourse.

Anyway, and thanks to the former chapter kernel's idea, some such uses can be approached by crisp sets that, nevertheless, cannot be seen as a fair representation of the corresponding word, but just as crisp approximations to it.

Obviously, the set $\{x \in [0, 10]; \mu_{small}(x) \leq \mu_{not\ small}(x)\}$ is the crisp kernel of negation of small, and its supremum separates what is qualifiable as small from what is not; hence there are some cases in which Black's separation point can be found. For instance, with the membership function $1 - x/10$, and the negation

1 − Id, the kernel is the interval [5, 10], and because the separation point is $s = 5$, a set "approaching" small in [0, 10] is [0, 5).

Of course, when available, the separation point depends on both the membership function (the word's meaning) and the negation; there is no "universal separation point" between small and not small, but one for each specification of small and not; it depends on how small and its negation are used in plain language.

For instance, were "small" used with the same membership function, but with the negation $N(x) = 1 - x/1 + x$, whose fixpoint is $(\sqrt{2} - 2)/4$, it would result in $s = 5/2(6 - \sqrt{12}) \approx 6.34$, with the kernel of negation [6.34, 10], and approaching set [0, 6.34). Approaching sets can be seen as crisp theoretic precisifications of the corresponding word, small in these cases. There is no single one, but many.

Of course, the point s will always exist provided the membership function is monotonic and defined in a compact set, but were it nonmonotonic, that is, for instance, decreasing in some part of X, and nondecreasing in another part, then s could not exist. Anyway, even if the membership function is asymptotic but monotonic, the separation point can exist.

For instance, with $\mu_{small}(x) = e^{-x}$ in $R^+ = [0, +\infty)$, and $N = 1 - \mathrm{Id}$, it is $e^{-x} \le 1 - e^{-x} \Leftrightarrow e^{-x} \le 0.5 \Leftrightarrow -\log_e 0.5 \approx 0.69 \le x$, and the set approaching this use of small in R^+ can be taken as the very short interval [0, 0.69), corresponding to the kernel of negation [0.69, +∞).

Of course, all that has been presented can be repeated with the crisp kernel of opposition $A(P)$, with which a larger approximation set may be obtained because $A(P) \subseteq N(P)$ implies $N(P)^c \subseteq A(P)^c$. Were P precise in X, and because, as was shown, $N(P)$ is the complement of the subset specified by P in X, the approximation set would coincide with that specified by P in X; what has been said reduces to what corresponds to the classical calculus. In each case, it should be previously decided if the separation point should be taken with the negation, or with an opposite word.

In conclusion, after the former numerical toy examples, it neither can be stated that Black's separation point always exists, nor that it can never be found. It is a question deserving more study and for which a finest acquaintance with the representations of the true uses of words in language seems to be unavoidable. Both controlled experimentation and mathematical modeling seem to be required.

20.3. Let's, provisionally, introduce the word "flat" to denote an opposite of heap. The correct use of the term *heap* requires the recognition that something is either flat, or is not a heap; in this way, and once a universe is fixed, the use of the word *heap* can be well learned, as with *small* in an interval of the real line, or with *odd* in the set of positive integers, and so on.

Under the typical hypothesis that a heap H only depends on the number N of grains it contains, and supposing N is a big enough integer, the universe of grains can be taken as the set of integers between 0 and N, and the degree up to which H is a heap described by the piecewise linear function:

$$\mu_H(x) = 0, \text{ if } 0 \leq x \leq N/3;$$
$$\mu_H(x) = 3x/N - 1, \text{ if } N/3 \leq x \leq 2N/3;$$
$$\mu_H(x) = 1, \text{ if } 2N/3 \leq x \leq N.$$

Then, with the negation 1 − Id, is:

$$\mu_H{}'(x) = 1 - \mu_H(x) = 1, \text{ if } 0 \leq x \leq N/3;$$
$$\mu_H{}'(x) = 2 - 3x/N, \text{ if } N/3 \leq x \leq 2N/3;$$
$$\mu_H{}'(x) = 0, \text{ if } 2N/3 \leq x \leq N.$$

And with the symmetry $N - x$, is:

$$\mu_H{}^a(x) = \mu_{\text{flat}}(x) = \mu_H(N - x) = \mu_H{}'(x),$$

showing that H^a = flat coincides with H' (not heap), in agreement with the inexistence of antonyms for heap, and making H^a (flat) appearing as a nonregular opposite.

Undoubtedly, a heap is constituted by grains although it is not perceived through the number of grains it contains, and that no observer will never try to count, but by its three-dimensional shape in which it should be a balance between the area of its base, and its height; for instance, with a small base and a large height, the heap will go down.

Considering that heaps are in three-dimensional Euclidean space, prototypes of heaps are, for instance, pyramids and circular cones; in the first case, because its volume is V = 1/3(base's area × height), the ratio or balance volume/height is V/h = base's area/3; in the second, because $V = 1/3(\pi \times r^2 \times \text{h})$, the ratio is $V/h = \pi \times r^2/3$. In both cases, the balance is one third of the area of the base; a ratio that seems possibly recognizable for a heap by a perceptive estimation.

With N a sufficiently large integer, denote by $S(p)$, $p \leq N$, a set of p grains, with which it can be stated "$S(p)$ is a heap"; in addition, the set $\{S(p); p \leq N\}$, can be endowed with the order $S(p) \leq^* S(q) \Leftrightarrow p \leq q$, and it can be recognized if $S(p)$ were a heap by perceptively comparing it with a prototype, such as it is a circular cone. For such a goal, let $\lambda(S(p)) \in (0, 1]$ a coefficient perceptively obtained by comparing the shape of $S(p)$ with that of a prototype, and only submitted to verify that if $S(p^*)$ has p^* grains, and if it is $p \leq p^*$, then $\lambda(S(p)) \leq \lambda(S(p^*))$.

For instance, were the prototype a circular cone with height 1 m, and base radius 1/2 m, whose volume is $\pi/12$ cubic meters, a presumed heap showing an estimated similarity of 50% ($\lambda = 1/2$), would have an estimated volume of around $\pi/48$ m^3. Of course, $\lambda(S(p)) = 0$ means that no similarity is perceived with the prototype, and $\lambda(S(p)) = 1$ that a full similarity is perceived.

Then, with an order automorphism f of the unit interval, it can be defined,

Degree up to which '$S(p)$ is a heap' = $\mu_H(S(p)) = \lambda(S(p)). f(p/N)$, because $S(p) \leq^* S(q) \Leftrightarrow p \leq q \Leftrightarrow p/N \leq q/N \Leftrightarrow f(p/N) \leq f(q/N)$, and as it should be also $\lambda(S(p)) \leq \lambda(S(p^*))$, it is $\lambda(S(p)).\ f(p/N) \leq \lambda(S(p^*)).\ f(q/N) \Leftrightarrow \mu_H(S(p)) \leq \mu_H(S(q))$.

Obviously, it is $\mu_H(S(p)) \in [0, 1]$, and by suitably choosing f, different models can be considered, such as the linear one with $f = \text{Id}$, $\mu_H(S(p)) = \lambda(S(p))p/N$, a quadratic model with $f(x) = x^2$, $\mu_H(S(p)) = \lambda(S(p))p^2/N^2$, and so on. Note that:

(1) With $p = N$, $\mu_H(S(N)) = \lambda(S(N))$, shows that $\lambda(S(N))$ is the degree up to which $S(N)$ is a heap.
(2) $\mu_H(S(p)) = 1 \Leftrightarrow \lambda(S(p)) = 1/f(p/N)$ that, in the linear model means λ (S (p)) = N/p.
(3) $\mu_H(S(p)) = 0 \Leftrightarrow \lambda(S(p)) = 0$, or $f(p/N) = 0 \Leftrightarrow \lambda(S(p)) = 0$, or $p = 0$.
(4) $\lambda(S(p)) = 1 \Rightarrow \mu_H(S(p)) = f(p/N)$, that if $f = \text{Id}$, would be just p/N.
(5) $\lambda(S(p)) = 0 \Rightarrow \mu_H(S(p)) = 0$.

Provided it is $\lambda(S(p)) > 0$, and that "$S(p)$ is not a heap" can be represented by a strong negation N_g, $\mu'_H = N_g \circ \mu_H$, then:

$$\mu'_H(S(p)) \geq \mu_H(S(p)) \Leftrightarrow g^{-1}(1 - g(\lambda(S(p))f(p/N))) \Leftrightarrow \lambda(S(p))f(p/N)$$
$$\Leftrightarrow g^{-1}(1/2)/\lambda(S(p)) \Leftrightarrow f(p/N) \Leftrightarrow Nf^{-1}(g^{-1}(1/2)/\lambda(S(p))) \geq p,$$

and the crisp kernel of negation is the interval $[0, Nf^{-1}(g^{-1}(1/2)/\lambda(S(p))]$. For instance,

- If $g = \text{Id}$, and $f(x) = x^2$, the separation point between heap and not heap would be $N\sqrt{(1/2\lambda(S(p)))} = 0.71N\sqrt{(1/\lambda(S(p))}$, that, with λ $(S(p)) = 1$, would be $0.71N$, and whose integer part could be taken as the number of grains.
- If $g = f = \text{Id}$, the separation point would be $N/(2\lambda(S(p))$, that, with $\lambda(S(p)) = 1$, would be $N/2$ grains, of which it suffices taking its integer part as the number of grains.

20.4. Up to some extent, analyzing sorites with just the help of classical Boolean reasoning recalls either studying nature before Renaissance times, or the heavens before the telescope. Anyway, this section is just an approach to the questions on sorites close to those a layperson could pose, and that, even accepting that a heap is constituted by grains, never will even try to count them.

As with everything in this book, such intent starts from a situational point of view considering the context in which things are perceived and concepts are learned, a naïve point of view that, nevertheless, allows going a little ahead in, for instance, the controversy on the possible existence of a separation point between what is and what is not, and its founding, by showing that even some imprecise words applied to a bounded numerical universe can show two such points, one for the kernel's complement of negation, and the other for the larger complement of the kernel of opposition. And, additionally, it opens the question of both numerical but not bounded universes and imprecise words not representable by monotonic membership functions. It is obvious that the topic still deserves further study.

Concerning the membership function of the term "heap", what is attempted is the introduction of a way of analyzing it by comparison with a prototype, but many aspects of it remain inconclusive; *heap* is a word that is well enough managed in language, but whose description is not easy.

In conclusion, this section tries to show that translating a conversation involving imprecise terms, such as heaps, into formal symbols, is still far from being clarified. Passing from perceptions and words to a calculus with them to reach conclusions is still further from the Leibniz wish of just computing. In particular, a realistic calculus with imprecise words whose meaning can be approached by crisp kernels of negation, or of opposition, is unknown, something that comes from the ignorance of which elements in the universe of discourse are approximately covered by words and its negations. It lacks, perhaps, a theory of "approximate crisp sets" of which, perhaps, the theory of rough sets is but an example.

20.5. What is really unknown is how to "approach by sets" the linguistic collectives generated by imprecise words not actually specifying a subset in the universe of discourse, a classification of them, and a realistic calculus with such approximations that can preserve the corresponding approximations.

A last question in such respect is whether such a calculus is possible without membership functions, but with sets coming from measures of the meaning of the corresponding words (such as they are the former kernels), and endowed with a suitable algebra. Something that, for what has been said, seems to be very doubtful with, at least, a unique algebra and, less again, if all of them are tried to be endowed with a lattice's structure. Anyway, and even vaguely, the question is posed.

References

1. E. Trillas, L.A. Urtubey, Towards the dissolution of the sorites paradox. J. Appl. Soft Comput. **4**(2), 1506–1510 (2011)
2. E. Trillas, I. García-Honrado, 20, A layperson reflection on sorites, in *Fuzziness in Medicine*, ed. by R. Seising et al. (Springer, Berlin), pp. 217–231
3. D. Graff, Phenomenal continua and the sorites. Mind **110**(440), 905–935 (2001)
4. M. Black, Vagueness: an exercise on logical analysis. Philos. Sci. **4**(4), 427–445 (1932)

Chapter 21
A Few Questions on Naming Concepts

Vagueness can be seen as coming from a lack of distinguishability between those instances falling into both a word and its negation, and creating a borderline zone with elements not clearly separable by negation, something coming from a lack of distinction, or distinguishability, from the meaning of a word and its negation. In what is possible it would be desirable to deal with the word "vague" when it is measurable, and produce a coverage of the elements in clusters, usually with nonempty intersections and that, at least for the borderline zones, seem to need to be seen as fuzzy sets.

21.1. Analogously as formerly described, typical instruments for reaching such a goal are those fuzzy relations $R: X \times X \to [0, 1]$, verifying the three laws:

(a) $R(x, x) \geq r$
(b) $R(x, y) = R(y, x)$
(c) $F(R(x, y), R(y, z)) \leq R(x, z)$,

for a fix $r \in (0, 1]$, all x, y, z in X, and called F–r-indistinguishability relations. The first law is r-reflexivity that coincides, when $r = 1$, with what has been denoted reflexivity, and then, if F is a t-norm T, R is just a T-indistinguishability relation, but $0 < r \leq 1$ allows us to keep some imprecision on the picturing of reflexivity by accepting a level of indistinguishability in recognizing each x, the second picture's symmetry, and the third T-transivitity. Notice that trouble appears from

$$R(x, y) \geq F(R(x, x), R(x, y)) = F(r, R(x, y)),$$

that provided $F(a, b) \leq b$, can imply that r should verify $R(x, y) = F(r, R(x, y))$ for all pairs x, y; for instance, were F = prod or F = min, it would imply $r = 1$, and with $F = W$ it is also $a = W(r, a) \Leftrightarrow a = \max(0, r + a - 1) \Leftrightarrow r = 1$. Hence, r-reflexivity forces us to define F-transitivity (c) with a function F different from a t-norm, or not to consider the universe's totality but a part of it.

E. Trillas, *On the Logos: A Naïve View on Ordinary Reasoning and Fuzzy Logic*, Studies in Fuzziness and Soft Computing 354, DOI 10.1007/978-3-319-56053-3_21

For instance, for some important relations such as those in $X = (0, 1]$ given by $R(x, y) = T(x, y)$, that are symmetric and min-transitive, $R(x, x) = T(x, x)$ is usually different from 1, but, for instance, it is prod $(x, x) = x^2 \geq r$, iff $\sqrt{r} \leq x$, $\min(x, x) = x \geq r$, and $W(x, x) = \max(0, 2x - 1) \geq r$, iff $x \geq (1 + r)/2$. Thus, the r-transitive law allows counting with a range in X, where a kind of reflexivity under a threshold can be considered. Where $r = 0.9$ (close to one), it is, respectively, $\sqrt{0.9} = 0.95$ and 0.9-reflexivity holds for prod between 0.95 and 1, for min between 0.9 and 1, and for W also between 0.95 and 1.

F–r-indistinguishability relations, when applied to $[0, 1]^X$, help to graduate the lack of distinction between a fuzzy set and its negation by the number $R(\mu, \mu')$, or the corresponding degree of distinguishability $1 - R(\mu, \mu')$ between them.

21.2. Let's make a remark on a very theoretical view. In a logical calculus introduced with logical "equivalence" ($\approx$), the famous Polish logician J. Lukasiewicz defined the *verum* of a statement p as the new statement $(p \approx p)$, and the *falsum* of p as $(p \approx 0)$, by denoting with 0 a "false" statement such as it can be $p \cdot p'$. Analogously, the *diffusum* of p can be defined as $(p \approx (p \approx 0))$, the equivalence between verum and falsum.

Given a T–r-indistinguishability relation R, $V(\mu) = R(\mu, \mu)$ can be defined as the verum of μ, the falsum as $F(\mu) = R(\mu, \mu_0)$, and the diffusum as $D(\mu) = R(\mu, R(\mu, \mu_0))$. With it, the function diffusum permits quantifying the indistinguishability of imprecise words whose membership functions or measures of their meaning are known. In some cases, the falsum coincides with the negation μ', and then $D(\mu) = R(\mu, \mu')$.

21.3. Some linguistic predicates are instances of words that are irreducibly imprecise in the corresponding context, and most of them are often learned in an "ostensive" way; for instance, teaching a child what the color red is, often is done by showing him or her some red objects. Thus, the labeling of new properties is frequently carried out by indistinguishability with other objects considered prototypical for such a label. In fuzzy logic and as formerly pointed out, there is the theoretically unsolved problem of labeling those membership functions resulting after operating some labeled membership functions, the so-called "linguistic approximation" problem. As formerly said, naming objects, or concepts, is always important.

Science is a process of constant invention, in which new horizons are continually viewed, and in which the problem of giving a name to something new is posed. Even if this can be debatable, it actually seems reasonable when referring to a scientific language; names of the newly appearing concepts are often coined due to some similarity they keep with other already known concepts, and even in a very different situation. It is in this context that the problem of naming a new concept can be posed and sometimes solved by synonymy. Notwithstanding, and in a large enough piece of language, synonymy cannot be seen as an "exact" phenomenon; what follows tries to explain this impossibility.

Let E be a universe of discourse, and $O = \{o_i\}$ a set of observers with a set T_0 containing linguistic terms, words t enabling the naming of some of the objects in X. The relation L_0 establishing a correspondence between objects in X and linguistic terms in T_0 is the "object language" of O. In the same way, the set $E(t) = \{x \in X; (x, t) \in L_0\}$ can be called the "extensional meaning" of the term $t \in T_0$, and it can be said that two terms t and t^* are synonyms in L_0, $t \approx t^*$, provided $E(t) = E(t^*)$; that is, if they have the same extensional meaning.

Nevertheless, this "precise conception" of synonymy in L_0 cannot capture the before-mentioned phenomenon of the breaking of synonyms' chains because relation $\approx$ is an equivalence, and because of its transitivity each chain $t_1 \approx t_2 \approx \cdots \approx t_n$ implies $t_1 \approx t_n$.

The problem is not of exact extensional meaning, but of meaning as a measure, that is, of representing each term t by a membership function μ_t of a fuzzy set labeled t. With it, a numerical degree of synonymy between terms can be defined by

$$S(t, t^*) = 1 - d(\mu_t, \mu_{t^*}),$$

with a distance d having sense between the "fuzzy extension" of terms, and representing how far t is from t^*. It is obviously $S(t, t) = 1$, and $S(t, t^*) = S(t^*, t)$, and it can also be easily proven that

$$W(S(t, t^*), S(t^*, t^{**})) \leq S(t, t^{**}), \text{ for any terms } t, t^*, t^{**},$$

thanks to the triangular inequality $d(t, t^{**}) \leq d(t, t^*) + d(t^*, t^{**})$ of the distance d. Therefore, once a suitable distance d is chosen, the index $S = 1 - d$, is a W-indistinguishability relation coming from the properties of d, and allowing, as seen formerly, capturing the breaking of synonymy chains.

For each suitable threshold of synonymy $\varepsilon > 0$, the approximate relation of synonymy can be set

$$t \approx_\varepsilon t^* \Leftrightarrow \varepsilon \leq S(t, t^*) \Leftrightarrow d(t, t^*) \leq 1 - \varepsilon,$$

that, additionally, contains the former particular case $\varepsilon = 1$ with $d(t, t^*) = 0$. This reflexive, symmetrical, but not transitive, precise relation can enable managing the breaking of the chains of synonyms, and allows approaching the problem of naming a new concept t by means of other already used names t^* and t^{**}. It shows how the indistinguishability can come from a distance between the linguistic terms.

The strategy consists in looking for linguistic terms slightly different from others already known, that is, in finding terms t^*, t^{**} such that satisfy, with S, the W-transitive inequality by verifying $d(t, t^*) \leq 1 - \varepsilon$, and $d(t^*, t^{**}) \leq 1 - \varepsilon$, with ε the selected threshold of synonymy. It is, perhaps, in this way that the speakers of a language can feel inclined to use one of these terms to name t; in this form an approximate synonym of t^* and t^{**} will be obtained. Associating a name with an object or concept, known as "binding" it, is actually relevant in many fields from science and technology, to management and business.

All that is, in addition, a way under which synonymy spreads vagueness throughout language, and shows that to represent it in measurable forms, membership functions of fuzzy sets are unavoidable. The relation of graded synonymy can allow us to approach the problem of naming new concepts by analogy, considering some of those already named, seen as close to them, leading to some economy of language, and allowing the intermingling of meaning that is necessary for reasoning.

In reasoning there is often a manifest incompatibility between precision and significance: too much precision can lead to an excessive shortening of the meaning of statements, and that can cut either a problem's fertile view, or the posing of good questions.

As Lotfi A. Zadeh wrote in respect to his "principle of incompatibility," "Stated informally, the essence of this principle is that as the complexity of a system increases, our ability to make precise and yet significant statements about its behavior diminishes until a threshold is reached beyond which precision and significance (or relevance) became almost mutually exclusive characteristics."

References

1. A. Sobrino, E. Trillas, Can fuzzy logic help to pose some problems in the philosophy of science? in *Representations of Scientific Rationality*, ed. by A. Ibarra, et altri (Rodopi, Amsterdam, 1997), pp. 277–300
2. L.A. Zadeh, Outline of a new approach to the analysis of complex systems and decision processes. IEEE Trans. Syst., Man and Cybern. **3**, 28–44 (1973)
3. M. Black, *Critical Thinking. An Introduction to Logic and Scientific Method* (Prentice-Hall, New York, 1946)
4. W.v.O. Quine, *Word and Object* (John Wiley & Sons, New York, 1960)
5. A.R.D. Prasad, N. Guha, Concept naming vs. concept categorization: a faceted approach to semantic annotation. Online Inf. Rev. **32**(4), 500–510 (2008)
6. J. Recasens, *Indistinguishability Operators* (Springer, Berlin, 2010)
7. E. Trillas, An approach to fuzziness in the setting of Lukasiewicz logic, Proc. IEEE Int. Symp. Multiple-Valued Logic, 222–226 (1983)
8. E. Trillas, E. Castiñeira, A. Pradera, On the equivalence between distances and T-indistinguishabilities, Proc. EUSFLAT-ESTYLF, 239–242 (1999)
9. E. Trillas, Apunte sobre la indistinguibilidad. Theoria **8**(1), 23–49 (1993)
10. E. Trillas, A.R. de Soto, On the thresholds of bounded pseudo-distances. Mathware Soft Comput. **15**, 189–200 (2008)
11. E. Trillas, Ll. Valverde, An inquiry into indistinguishability operators, in *Aspects of Vagueness*, ed. by H. Skala, et altri (Reidel, Dordrecht), pp. 231–256

Chapter 22
Instead of a Conclusion

It is difficult to conclude something when almost no certitudes, but mainly reflections, questions, and doubts, are actually presented, and most of whose answers are, actually, still "blowing in the wind."

Such reflections leading to questions and doubts appeared, especially, while Lotfi A. Zadeh was introducing his, initially academically heterodox, ideas on the new field of "computing with words" (CwW), a challenging and exciting one in which the potentiality of the creative ideas it contains can truly flourish, and whose basis I see grounded on ordinary reasoning and plain language.

It may be that the only conclusion for this not very technical book, is that CwW = fuzzy logic, cannot be further seen as just a theoretic and formal discipline, that it should be rebuilt as a discipline of an experimental type with formal models testable against the reality of plain language and ordinary reasoning. Such formal models are necessary, for instance, for the design of linguistically described systems.

As it can be said that matter and energy are the grounding of physics, plain language and ordinary reasoning are the grounding of fuzzy logic. The plain reasoning people do should not be confounded with the formal reasoning that is done, let's say, with pencil and paper thanks to some mathematical framework. The second is but a more or less trustworthy translation of the first that, notwithstanding, benefits from the calculus the framework facilitates.

For completing the ideas that generated the book there still lack:

- Explaining how its author saw (of course, in a subjective personal form), the prolegomena to the introduction of CwW in 1993, and what motivated the presented approaches to ordinary reasoning and the linguistic meaning of words.
- Some additional comments on classification, and on the modalities on which modus ponens (MP) appears.

22.1. When Charles A. Elkan published his controversial paper in 1993 on the "paradoxical success" of fuzzy logic, many researchers among the most qualified in

E. Trillas, *On the Logos: A Naïve View on Ordinary Reasoning and Fuzzy Logic*,
Studies in Fuzziness and Soft Computing 354, DOI 10.1007/978-3-319-56053-3_22

the "fuzzy" field, replied to him in papers in which they responded by contradicting what he wrote.

Elkan's paper begins with a theorem proving that, in the standard algebra given by the triplet (min, max, 1 − id), a truth degree t with values in [0, 1] reduces to only values in {0, 1}, provided some logical form holds. It is based on the analysis in such algebra of the formula $p' + q = q + p' \cdot q'$, the equality between the typical Boolean conditional and the Dishkant one used in quantum logic that, as is known, are identical in Boolean algebras.

The structure given in [0, 1] by the triplet (min, max, 1 − id) is not Boolean, but a De Morgan algebra with Boolean elements 0 and 1, in which the distributive law $q + p' \cdot q' = (q + p') \cdot (q + q')$ holds, and without always being $q + q' = 1$, it is not possible to conclude the former equality for all p and q, but only the obvious inequality $q + p' \cdot q' \leq q + p' = p' + q$.

Elkan's theorem is neither surprising nor destructive for fuzzy logic; its argument can be repeated, and more easily, with formulae such as $p + p' = 1$, with which $1 = t(p + p') = \max(t(p), 1 - t(p))$, implies $t(p) \in \{0, 1\}$. It is even not difficult to prove that the formula used by Elkan is equivalent, in any De Morgan algebra, to the formula $q \cdot q' \leq p$ that, with $p = 0$, shows it is only valid if $q \cdot q' = 0$, that is, if q is a Boolean element, if either $q = 0$ or $q = 1$. Hence in ($[0, 1]^X$, min, max, 1 − id) the formula only holds with the crisp sets.

The only thing that such a theorem shows is that in De Morgan algebras there are Boolean formulae, classical laws, not holding universally, something already well known in 1993. The debate around Elkan's paper even induced a paper on some sociological aspects of scientific research in which the true spirit of theoretic fuzzy logic is not present; up to some extent, the polemics surpassed the limits of AI and fuzzy logic.

Some years later, around 2000, and jointly with Claudi Alsina, we studied in which standard algebras of fuzzy sets ($[0, 1]^X$, T, S, N), the formula $\mu' + \sigma = \sigma + \mu' \cdot \sigma'$ can hold, and, by solving the corresponding numerical functional equation, $S(N(a), b) = S(b, T(N(a), N(b)))$, we obtained that these algebras are those given by $T = \text{prod}_f$, $S = W_f^*$, and $N = N_f$ and, consequently, that such a law cannot coexist with the duality laws in, at least, the standard algebras of fuzzy sets, thus, that in plain language there could be some parts of it showing imprecision, in which, without duality, the formula used by Elkan can actually hold.

Because the previous formula does not reflect a short statement, we also studied the case in which the "or", and the "not" appearing in it, are not represented by the same S and N; that is, studying the more complex functional equation $S_1(N_1(a), b) = S_2(b, T(N_2(a), N_3(b)))$, from which it easily follows that $N_1 = N_2 = N_3$, and many solutions with different S_1 and S_2 were found.

Finally, we also studied where von Neumann's law of perfect repartition $\mu = \mu \cdot \sigma + \mu \cdot \sigma'$ can hold, whose validity, as formerly explained, also excludes duality. Were all the Boolean laws valid, a standard algebra would be a Boolean algebra and, as previously shown, no BAF (basic fuzzy algebra) can be an ortholattice.

In the end, all that does not represent something surprising in fuzzy logic, but opened my eyes towards reflecting on the minimal number of laws necessary for trying to establish an initial formalization of ordinary reasoning based on plain language, and, at least, within environments of imprecision. Under this view, I established the axioms for a BAF that, with a new abstraction step, can still be weakened as follows.

22.2. A quadruplet $(X, <; '; \cdot, +)$, with $X = \{p, q, r, \ldots\}$, verifying the axioms,

(1) $p<q \Rightarrow q'<p'$
(2) $p \cdot q<p$, and $p \cdot q<q$
(3) $p<p+q$, and $p+q<q$
(4) There exists a nonempty proper subset B of X, for all whose pairs of elements (p, q), there holds $p \cdot p' < q + q'$.

can be called a basic abstract algebra (BAA), and, of course, either ortholattices (and a fortiori orthomodular lattices and Boolean algebras), De Morgan–Kleene algebras, and BAF, are BAAs.

A basic abstract algebra is but a very weak algebraic structure whose first three axioms are just those allowing us to pose the representation of ordinary reasoning in a primitive, symbolic, and naïve form, although for either studying some parts of reasoning or reaching some results some additional axioms on negation ('), conjunction ($\cdot$), and disjunction (+), as well as on the transitivity of <, seem to be needed and, consequently, added. The failure of such conditions, as well as how p and $(p')'$ are either linked by <, or are not, seem to be responsible for a big chasm between the reasoning formalized in strong structures, and some important aspects of ordinary reasoning, such as creative reasoning.

Hence, it seems that the BAA algebraic structure is a minimal one that, by sometimes adding more axioms to it, can permit some advances towards a formal symbolic analysis of the complex phenomena of ordinary reasoning. We did it by showing that it only consists in conjectures and refutations, that speculations can facilitate hypotheses, how to falsify hypotheses, and so on. Indeed, everything that is presented in Chap. 8 can be formalized in a BAA.

To continue such study, more empirical knowledge on ordinary reasoning is required; that is, some parts of nondeductive ordinary reasoning should be submitted to processes of not blind but systematic observation, followed first by the establishment of mathematical models, then by controlled experiments trying to falsify such models, provisionally accepting a model, and so on. This path is but the usual, always unending, for acquiring scientific knowledge. It could be initiated, for instance, by experimentally testing some of the results presented in this book's Part I, Sowing Ideas; until it is done, these results will remain but working hypotheses.

Nevertheless, the only researcher who offered a new view, after the polemics provoked by the Elkan paper, was Zadeh, whose not too much farther in time proposition of fuzzy logic as CwW, returned to the same origin of fuzzy sets, and simultaneously opened a new way towards trying to solve the complex practical problems posed by the representation of questions described in plain language, and the ordinary reasoning people employ for dealing with them, towards copying with the reasoning people do. This was something that could mean a new step in the long path Leibniz started with his famous shout, "*Calculemus!*" for which a calculus with words is essential, and that cannot be thought only within the framework given by the former four axioms of a BAA, but requires the possibility of extending them with more suitable axioms at each particular case.

What should be avoided is trying to study ordinary reasoning theoretically with either a single mathematical framework, or presuming laws without being sure of their actual validity. It does not seem that a single mathematical model can represent "all" ordinary reasoning; jointly with plain language, ordinary reasoning constitutes a system that is, perhaps, the most complex system with which computer science is currently faced. Both language and reasoning are essentially intermingled, and the second cannot be deeply studied without knowing more about the first.

Zadeh's CwW represents a big theoretical challenge that, in my view, requires looking at plain language and ordinary reasoning from broader points of view than those currently assumed by just mathematical fuzzy logic, something for which more knowledge is necessary. For instance, and to quote just two actually important and lacking topics, neither a mathematical model of ambiguity is currently known, nor is there general agreement on how to deal with uncertainty and probability, in particular for nonrandom and imprecise events as these concepts appear in plain language.

22.3. Concerning the model of meaning as a quantity, it came from the Aldo De Luca and Settimo Termini idea of a nonprobabilistic "fuzzy entropy", that is, on the measuring of fuzziness, with which this concept appeared as a strengthening of the wider philosophical concept of vagueness, namely as a measurable part of it. With their fuzzy entropies, De Luca and Termini reached a new view of vagueness that, although referring to only a part of it, meant an advance towards its still incomplete scientific domestication.

The basic problem for measuring vagueness lies in the difficulties appearing for counting with a clear form to recognize when "this is less vague than that", in part due to the appearance of vagueness in a multitude of different contexts, with intermingled situations in a given context and coming from the different sources from which vagueness arises, that is, in a particular universe of discourse X, founding a graph $(X[P], <_{\text{vague}})$, with "vague" the mother-predicate of the concept "vagueness" and, at least, applied to statements "x is P", in short, for capturing the qualitative meaning of the word "vague" even in a relatively simplified context. This is just what, changing vagueness by fuzziness, De Luca and Termini did with the predicate "fuzzy" applied to those words that are represented by membership functions in X.

De Luca and Termini translated "P is less fuzzy than Q" into "μ_P is less fuzzy than μ_Q" once, obviously, the meanings of P and Q in X were, respectively, measured by their (designed) membership functions. With this in mind, they presumed

$\mu_P <_{\text{fuzzy}} \mu_Q \Leftrightarrow$ The values $\mu_P(x)$ are closer to 0 and 1 than those of $\mu_Q(x)$ are, and suggested them to define the relation $<_{\text{fuzzy}}$ by

$$\mu_P <_{\text{fuzzy}} \mu_Q \Leftrightarrow 0 \leq \mu_P(x) \leq \mu_Q(x) \leq 0.5, \text{ or } 0.5 \leq \mu_Q(x) \leq \mu_P(x) \leq 1.$$

Note that, indeed, what they actually did was to change the linearly ordered unit interval ([0, 1], $\leq$) by the same unit interval [0, 1] but endow it with the sharpened partial order $\leq_s$, defined by

$$a \leq_s b \Leftrightarrow \text{Either } 0 \leq a \leq b \leq 0.5, \text{ or } 0.5 \leq b \leq a \leq 1,$$

splitting the unit interval, and pointwise translating it into $[0, 1]^X$. Hence, because the sharpened unit interval ([0, 1], $\leq_s$) has the two minimals 0 and 1, and the maximum 1/2, in $[0, 1]^X$ the minimals with respect to $<_{\text{fuzzy}}$ are just the membership functions of the crisp sets, and there is only a maximal, its maximum, the membership function $\mu_{1/2}$ constantly equal to one half.

Thus, once the graph ($[0, 1]^X$, $<_{\text{fuzzy}}$) is established representing the qualitative meaning of the predicative word "fuzzy", a measure of this word's meaning is but a mapping E: $[0, 1]^X \to [0, 1]$ such that:

(1) $\mu <_{\text{fuzzy}} \sigma \Rightarrow E(\mu) \leq E(\sigma)$
(2) $\alpha \in \{0, 1\}^X \Rightarrow E(\alpha) = 0$
(3) $E(\mu_{1/2}) = 1$

a definition that is coincidental with that of a "fuzzy entropy" given by De Luca and Termini by only changing $<_{\text{fuzzy}}$ by $\leq_s$.

Hence, after accepting that the use, or qualitative meaning, the predicate "fuzzy" shows in $[0, 1]^X$ can be recognized once reflected by the partial order relation $<_{\text{fuzzy}} = \leq_s$, each measure E facilitates a full meaning of "fuzzy" by means of the corresponding quantity ($[0, 1]^X$, $<_{\text{fuzzy}}$, E).

As always, there is not a unique E; in each case, to specify one of such measures, more conditions should be presumed according to the corresponding context, and, after the paper by De Luca and Termini was published, several papers were devoted to different possible models for the measure E. In addition, and with all that, the concept of fuzziness acquired a scientific context with which vagueness is not yet endowed. Since then, fuzziness can be seen as a scientifically domesticated part of vagueness; it appears as a type of (measurable) vagueness restricted to words with a measurable meaning, permitting us to take into account how much fuzzy/vague a word is in language and, eventually, trying to reduce or, at least, numerically control its fuzziness. In addition, and when representing P, between the several possible membership functions, it seems suitable to select one with minimal entropy; that is, choosing the less than possible fuzzy one. Fuzziness is often not avoidable, but it is always better to work with membership functions showing it as less as possible; in no case is clarity to be disdained.

Anyway, and although the idea of entropy is presented here in terms slightly different from those De Luca and Termini originally used to introduce it, their ideas are what led me to see meaning as a quantity; allowed me to see the "ideal" membership function of a fuzzy set as a measure of its linguistic label's meaning; helped me to pose the meanings of the linguistic particles, or connectives, *and*, *or*, *not*, introducing an analysis of the opposites, as well as their difference with negation; and so on.

Note that De Luca and Termini took the threshold of fuzziness as 1/2, which is just the fixpoint of the negation $1 - \text{id}$, but, provided negation should be represented by a different strong negation N, then the laws of a fuzzy entropy can be slightly modified by just taking the fixpoint of N instead of 1/2. It also could be thought of as substituting negation by an opposite, an option that remains an open possibility if the used negation is not strong.

As with the polemics of Elkan and Zadeh's introduction of CwW, which led to reflecting on the necessity of going towards a symbolic study of ordinary reasoning, the De Luca–Termini trial studying the linguistic label "fuzzy" led to seeing the meaning of predicative words as a quantity, also by abstracting the general concept of measure from Michio Sugeno's [11] fuzzy measures. Reflecting on what wise people wrote is always very important; trying to advance from "the shoulders of giants" is always good.

22.4. Some last comments on the main topics touched on in this book can be pertinent. First and among them are those concerning the (practical) dangerous risk of using mathematical models endowed with laws or axioms whose practical validity has not been previously tested in the corresponding language's context. Second is the necessity of classifying the elements in the universe, that is, of separating them in what is possible for looking more clearly at them with those concepts taken into account, and that, in precise reasoning, is done by using crisp partitions coming from the perfect classification a set and its complement facilitate, something that does not happen, in general, with fuzzy sets labeled with imprecise words.

In the first respect, it is worthwhile explaining a true story. A colleague, and good friend of mine, came to me perplexed because the mathematical model he was using did not fit the contextual data of which he was totally sure. Because he knew that the conjunction was given by the *t*-norm prod, and because the distributive law $\mu + \sigma \cdot \lambda = (\mu + \sigma) \cdot (\mu + \lambda)$ holds, the problem appeared by believing that it forces the pair of connectives (min, max). The *t*-conorm max was all right with him, but $T = \min$ instead of $T = \text{prod}$ was what had him worried. Five minutes sufficed to convince him that there was no trouble at all, because such distributive law holds if and only if $S = \max$, regardless of which T can be taken.

This story shows that those designing fuzzy systems should be well acquainted with the theoretical armamentarium fuzzy logic counts with at each moment, and that if the adopted model does not fit the contextual information, the obtained solutions can be wrong. This is what would happen by substituting prod by min in my colleague's case, and illustrates a true risk that could come from using an either not well understood, or wrong, mathematical model. Designers of fuzzy systems should be acquainted with theoretical fuzzy logic.

22.5. Regarding classifications, they are needed for separating the universe in subcollectives responding to several linguistic labels, and that, often and later on, still could deserve to be refined in more subcollectives arising from the former ones.

With precise words, that is, with crisp sets, the problem consists in perfectly classifying the universe X in nonoverlapping subsets $A_1, \ldots, A_n$, such that their union is X; hence, for instance, the complement of A_1 is $A_2 U \ldots UA_n$, and it corresponds to generate a finite Boolean algebra by means of the n atoms $a_1, \ldots, a_n$, corresponding to the "names", or linguistic labels, specifying the subsets A_i.

The problem can come from an initial, precise, linguistic label P, reflecting a concept that is, later on and in its turn, classified by means of several predicates, such as in the case with conjectures that were classified in consequences, hypotheses, and speculations that, in turn, become classified in type-one and type-two speculations.

Note that a perfect classification $\{A_1, \ldots, A_n\}$ of X, comes from a classical equivalence relation having the subsets A_i as its classes of equivalence, that is, from a reflexive, symmetrical, and transitive previous relation.

Nevertheless, when the linguistic labels are imprecise, this is, in general, not possible, and a usual resource is given by the formerly introduced T-indistinguishability relations, functionally expressed through functions R: $[0, 1] \times [0, 1] \to [0, 1]$, verifying the reflexive, symmetrical, and T-transitive laws. Once a is fixed in $[0, 1]$, the fuzzy set with membership function $A(x) = R(a, x)$, for all x, can be seen as an imprecise class of T-indistinguishability, and, provided R only takes its values in $\{0, 1\}$, it is obvious that the former laws reduce to those of a classical equivalence and that functions A are but the characteristic functions of their (crisp) equivalence classes.

T-indistinguishabilities generalize crisp equivalences, reducing to them when their values just range in $\{0, 1\} \subseteq [0, 1]$. As formerly said, all definitions done in the "fuzzy field," should reduce to the classical ones for crisp sets, a reduction that is necessary for allowing any fuzzy calculus to deal with both precise and imprecise words because, indeed, statements can usually contain precise and imprecise words. Note also that $T(R(a, x), R(x, b)) = T(A(x), R(x, b) \leq A(b)$, means that the "fuzzy classes" A are but MP's states of R.

If R were not precise, and in any case, provided the set $\{(a, b); R(a, b) = 1\}$ were not empty (indeed, it would not be if R were reflexive), it would be a crisp relation that, obviously, is always a classical equivalence. What is different concerns the other "levels of crispness" of R, that is, the crisp sets in $X \times X$,

$$R(\varepsilon) = \{(a, b);\ R(a, b) \geq \varepsilon\},\ \text{with } \varepsilon \text{ in } [0, 1].$$

These crisp relations $R(\varepsilon)$ are, for $\varepsilon < 1$, always reflexive and symmetrical, but only are transitive if and only if R were min-transitive. Hence, only the fuzzy min-equivalences are decomposable in the crisp equivalences $R(\varepsilon)$, each facilitating a crisp partition of X. Because $\varepsilon \geq \tau$ implies '$R(a, b) \geq \varepsilon \Rightarrow R(a, b) \geq \tau$', the partition associated to $R(\tau)$ is a refinement of that associated with $R(\varepsilon)$.

In this form, given X and a fuzzy min-equivalence R, a triplet of crisp partitions of X is obtained, each indexed with a value of ε, and whose respective number of equivalence classes decreases from the level $\varepsilon = 1$ up to the level with the smallest value for ε.

For instance, the very simple fuzzy relation in the set $X = \{1, 2, 3\}$ given by

$$R(1, 1) = R(2, 2) = R(3, 3) = 1;$$
$$R(1, 2) = R(2, 1) = R(1, 3) = R(3, 1) = 0.5;$$
$$R(2, 3) = R(3, 2) = 0.7,$$

generates the three partitions of X,

$\{\{1\}, \{2\}, \{3\}\}$, with index $\varepsilon = 1$,
$\{\{1\}, \{2, 3\}\}$, with index $\varepsilon = 0.7$, and
$\{\{1, 2, 3\}\} = \{X\}$, with index $\varepsilon = 0.5$,

the first refining the second, and this refining the third. Note that such sequence of partitions goes from separating, or distinguishing, each element in X at the maximum level $\varepsilon = 1$, to separating nothing, that is, only distinguishing X at the minimum level $\varepsilon = 0.5$, and separating the subsets $\{1\}$ and $\{2, 3\}$ at the intermediate level $\varepsilon = 0.7$. Of these partitions, only the second seems actually informative, inasmuch as the first and the third just mean recognizing that there is a set X with different elements, but the second recognizes that elements 2 and 3 share something in common that separates them from element 1.

The situation is very different when departing from a predicative imprecise word. As was said, the classes are fuzzy sets whose overlapping is necessarily different from the crisp case because each one affects, in principle, all points in the universe, and only fuzzy clusters can appear.

An example of it lies in Zadeh's idea of a "linguistic variable", as it is with the imprecise predicate P = cold in a scale X of external temperatures. In it, the play among P, P^a = hot, and P^m = warm, suffices for obtaining, not obviously a crisp partition, but a fuzzy coverage of X. With it, each of the points (x) in the scale X counts with the associated triplet of numbers $(\mu_{\text{cold}}(x), \mu_{\text{warm}}(x), \mu_{\text{hot}}(x))$ that, provided their addition were up to 1 at each x, could be seen as a "fuzzy partition of X". Remember, in this respect, that in the crisp case each x in X belongs to just one of the classes; hence, the values such x takes for its membership in the n classes, are 0 for $n - 1$ of them, and 1 for that to which it belongs; that is, the sum total of these values is 1.

A linguistic variable arises from a linguistic label P, and is linguistically generated by P and P^a, or P', jointly with some more terms such "middle", or "very P", "more or less P", and so on. Hence a linguistic variable generated by P is but the set of words $\{P, P^a, P', P^m, vP, \ldots\}$ with which a (flexible) coverage of the universe related to the principal term P is obtained, and is known as a "fuzzy partition" provided the sum of the corresponding values for the respective linguistic labels is, at each point, equal to one.

The linguistic variable's idea is anchored in plain language where, to "better separate" the universe to which an imprecise predicate P is applied, between five and nine linguistic terms derived from P are often used, the so-called "seven plus or minus two" empirical law that was statistically established as the limits between which people are usually able to discriminate, or to graduate, or to internally separate, concepts. It should be noted that in the simpler applications of the calculus with linguistic variables, only three terms are often used.

The introduction of linguistic variables by Zadeh can be signaled as the first step from fuzzy logic to its concretion as CwW, because they are intrinsically related to plain language and to how people actually manage linguistic concepts. For instance,

- The linguistic variable "Age", {young, old, middle-aged, not old, not very young, …}
- The linguistic variable "Size", {large, small, medium, very large, very short, not very large, …}
- The linguistic variable "Truth", {true, false, very true, not very false, not true, …}
- The linguistic variable "Height" (for people), {tall, short, medium, very tall, not very short, …}
- The linguistic variable "Height"(for buildings, etc.), {high, low, medium, very high, not very high, not low, …}
- The linguistic variable "Speed", {fast, slow, not fast, very slow, not very fast, not slow, …}
- And so on

The given-name attributed to each linguistic variable corresponds to a concept of which either P or P^a is its mother-predicate; once the predicates actually used as the linguistic values of a linguistic variable are designed by membership functions in the universe X, theoretic fuzzy logic facilitates a calculus with them.

To end this section, let's point out an important difference the calculus with linguistic variables shows, in contrast with that of the classical (bivaluate) calculus of predicates; such difference is that, in the imprecise case, everything should be designed according to the available contextual information. Being as best as possible acquainted with how the imprecise predicates and the connectives are contextually used (i.e., their respective meanings in the corresponding universes of discourse) is fundamental for representing something well in fuzzy terms.

22.6. Provided M were a function between either $\{0, 1\}^X$ or $[0, 1]^X$ into $[0, 1]$ (e.g., the identity, a fuzzy set, a measure of probability or possibility, etc.), the MP inequalities would appear in one of the two forms,

- $T(M(p), R(p, q)) = M(p \cdot q) \le M(q)$,
- $T(M(p), R(p, q)) = T^*(M(p), M(q)) \le M(q)$,

provided the function M in the first inequality (the "internal" case), and the operation T^* in the second (the "external" case) permits reaching the inequalities. The names *internal* and *external* come from the two conjunctions before the symbol $\leq$ of inequality; the first is in the domain of M and between elements, and the second is in the range of M and between its images.

For instance, as has been shown, in Boolean algebras and in probabilized Boolean algebras, the MP inequality is attained in the forms:

- $p \cdot (p' + q) = p \cdot q \leq q$

and,

- $\text{prob}\,(p) \cdot \text{prob}\,(q/p) = \text{prob}\,(p \cdot q) \leq \text{prob}\,(q)$,

respectively. Also, for instance, in fuzzy logic the MP inequality is analogously reached in well-known cases including

- If $R(a, b) = \max(1 - a, b)$, is $W(a, R(a, b)) = W(a, b) \leq b$
- If $R(a, b) = 1 - a + a{\cdot}b$, is $W(a, R(a, b)) = a{\cdot}b \leq \text{b}$
- If $R(a, b) = \min(1, 1 - a + b)$, is $W(a, R(a, b)) = \min(a, b) \leq b$.

Such two routes for reaching the MP inequality can be expressed by the diagrams

- Internal, $p \rightarrow q, p \in V{:}p \cdot q \in V{::}q \in V$,
- External, $p \rightarrow q, p \in V{:}p \in V$ and $q \in V{::}q \in V$.

Hence similarly to the before-considered nonmonotonic relations, if the classical equivalence between "$p \cdot q \in V$" (p *and* q is V), and "$p \in V$ and $\in V$" (p is V *and* q is V), were broken, two different routes would actually exist for attaining the rule of modus ponens. As was said, the inference with nonmonotonic relations deserves more analysis.

Returning to plain language, it should be known if, when "q is V" does not follow from "p *and* q is V", it would hold that "q is V" follows from "p is V *and* q is V" if between the two "and", actually were or were not, the usually presumed equivalence. The different uses of the linguistic particle "and" are not yet fully analyzed, and, for instance, a way of looking at them not only as "joining" statements is still open, but as defining either forward or backward inferences by conditionals that can, or not, be monotonic.

References

1. E. Trillas, C. Alsina, Elkan's theoretical argument reconsidered. Int. J. Approximate Reasoning **26**, 145–152 (2001)
2. A. Pradera, E. Trillas, C. Moraga, Clarifying Elkan's theoretical result, Proc. EUSFLAT, 604–608 (2003)
3. Ch. Elkan, The paradoxical success of fuzzy logic. IEEE Expert **9**, 3–8 (1994)
4. IEEE Expert, August 1994, Issue 4

5. C. Rosental, Certifying knowledge: the sociology of a logical theorem in artificial intelligence. Am. Sociol. Rev. **68**, 623–644 (2003)
6. A. De Luca, S. Termini, A definition of a nonprobabilistic entropy in the setting of fuzzy sets theory. Inf. Control **20**(4), 301–312 (1972)
7. E. Trillas, A scrutiny on globally measuring imprecision, in *Liber Amicorum*, ed. by P. Greco, et al. (Doppiavoce, Naples, 2015), pp. 111–118
8. L.A. Zadeh, *Computing with Words* (Springer, Heidelberg, 2012)
9. E. Trillas, C. Alsina, A reflection on what is a membership function. Mathware & Soft Comput. **6**, 201–215 (1999)
10. E. Trillas, T. Riera, Entropies in finite fuzzy sets. Inf. Sci. **15**, 159–168 (1978)
11. M. Sugeno, Theory of fuzzy integrals and its applications, Ph.D. Thesis, Tokio Institute of Technology, 1974
12. S. Termini, On some vagaries on vagueness and information. Ann. Math. Art. Intell. **35**(1–4), 343–355 (2002)
13. L.A. Zadeh, The concept of a linguistic variable and its applications to approximate reasoning-Part I. Inf. Sci. **8**, 189–214 (1975)
14. G.A. Miller, The magical number seven, plus or minus two. Some limits in our capacity for processing information. Psychol. Rev. **63**(2), 81–87 (1972)
15. E. Renedo, E. Trillas, C. Alsina, On three laws typical of booleanity, Proc. NAFIPS, 120–123 (2000)
16. L.A. Zadeh, Fuzzy logic = computing with words. IEEE Trans. Fuzzy Syst. **4**(2), 109–111 (1996)

Chapter 23
To Conclude

The book has arrived at its end, and the following five quotations, always motivating the author, can allow us to imagine the roots of what it contains.

- The novelist Julio Cortázar was obsessed, in his youth, by looking at a dictionary, and his father asked him, "What do you search for in the dictionary?" He answered, "For all."
- Albert Einstein said that, "The chemical analysis of a soup has no taste."
- The Nobel Laureate physicist Isidor J. Rabi, when asked what made him a scientist, answered, "In my neighborhood, mothers used to ask their children if they did learn something new at School, but my mother always asked me if I did ask *good questions*. Even without intending it, she did make me a scientist."
- The writer Stefan Zweig wrote, in a known memoir, "The last pending mystery concerns people's creativity."
- The mathematician Karl Menger, in addition to recommending more "*exact thinking*" than possible, wrote that good questions are those allowing us to get *fertile answers*.

Brain and its functioning, language and reasoning, are, jointly with the senses, the arms and the hands, the legs and the feet, a good part of what Nature gifted us with for living. Of them, only language and reasoning can be seen as typically privative of the species Homo; the others are not the best among animals, not even mammals in particular. Imagining a person unable to express himself or herself by some type of language (oral, written, gestured, etc.) is very difficult, if not impossible. Today, for instance, computers can be managed by people without hands; information and communication are necessary for reasoning, and reasoning for living. Exact thinking is necessary for understanding.

Research (and not only the scientific one) requires people being endowed with curiosity, a taste for personal endeavor, intellectual skills for posing good questions, and specialized knowledge for getting fertile answers. For arriving at fertile answers it is necessary to be creative, and creativity requires large doses of curiosity and

E. Trillas, *On the Logos: A Naïve View on Ordinary Reasoning and Fuzzy Logic*,
Studies in Fuzziness and Soft Computing 354, DOI 10.1007/978-3-319-56053-3_23

imagination, hard study, and critical knowledge. Usually, neither good questions, nor its fertile answers, are unique; often, creativity comes from a skewed look at a problem.

Today science is in front of new frontiers for understanding the joint play of language and reasoning, and it counts with chances enough for posing good questions allowing fertile answers in such respect for trying to domesticate reasoning scientifically.

For it, the purely abstract point of view should be changed through a new trial towards an experimental science, dealing with the intermingled pair constituted by plain language and ordinary reasoning, a change towards a kind of physics of language and reasoning. For it, and because language and reasoning are strongly permeated by imprecision, a rethought and down-to-top new view of fuzzy logic surely would be relevant if and when it would not be further seen as a purely mathematical subject, and is completed with studying ambiguity, and how the uncertainty of nonrandom and imprecise "events" can become effectively measurable.

In addition to the previous (more insinuated between the lines than clearly posed) questions, the most currently challenging scientific and pending question is that referring to understanding what is actually the wellspring of progress, that is, creative reasoning. How it works, how monotony fails in it, how it is boosted by analogy, how it can be classified, and how it could be either computationally mechanized or truly known regarding its impossibility, are among the more important intellectual challenges. New forms of posing such questions seem to be needed.

In the end, "*Quidquit recipitur ad modum recipientis recipitur,*"('Whatever is received into something is received in the manner of the receiver', that can be paraphrased as 'What is in a container, takes the container's form') the Thomist (and static) principle so dear to the Middle Age's scholastic philosophers, often requires breaking down the old container for advancing towards something new. It happened, for instance, when the container with the real numbers did not allow capturing something new, and a larger container with complex numbers was introduced, not only new mathematical questions arose, but some aspects of physics were measured and rethought.

Neither thinking, nor knowledge, nor science, nor technology, are static, but dynamic, and now a breakthrough seems to be possible for scientifically domesticating creative reasoning or, at least, for knowing if it is not totally possible, and, in such case, up to what extent it is possible.

In the (literary) words of the Spanish philosopher José Ortega y Gasset, "Metaphor is one of the more fruitful potentialities of man. Its efficiency scraps magic, and seems to be an instrument for creation God did forget inside one of its creations."

It is hidden that creative reasoning indeed scraps magic by not being deductive and facilitating creativity; it is currently a big and intriguing mystery that, in my view, only would be theoretically clarified by researchers looking at the problem with a new and open look, including a naïve attitude and a personal feeling of how tasteful such a challenge is for them.

The reflections this book contains, and maybe just consists in, are written for calling the attention of those researchers who would try to afford the presented goals and questions by, but not only, breaking down the old container and changing it to a new one.

All in the book is, to some extent, endowed with some philosophical flavor, along the lines of Jerry A. Fodor's words in 1981, "The form of a philosophical theory, often enough is: *Let's try looking over here*."

This is not, indeed, a book directed to those who are already "specialists," but to researchers aiming at scrutiny into what is over there in the fog of plain language and ordinary reasoning.

In its deep background, the book is concerned with a trial for pragmatically approaching the human Logos that, in the praxis and essentially, consists in the capabilities of questioning, guessing, telling, and computing. It is, perhaps, from a pragmatic view of Logos that a methodology permitting mechanizing it up to the greatest degree possible can flourish. Anyway, agreeing with the Socratic, "All I know is that I know nothing," that what is unknown is much more than what is believed to be known, is always essential.

Complementary References

1. A. Aliseda, *Abductive Reasoning* (Springer, Dordrecht, 2006)
2. J.L. Austin, *Philosophical Papers* (Oxford University Press, Oxford, 1979)
3. J.D. Barrow, *Impossibility* (Oxford University Press, Oxford, 1998)
4. G. Bachelard, *L'engagement rationaliste* (Presses Universitaires de France, Paris, 1972)
5. G. Bachelard, *Essai sur la connaisance apprroché* (Vrin, Paris, 1981)
6. J.L. Bell, M. Machover, *A Course in Mathematical Logic* (North Holland, Amsterdam, 1997)
7. E. Beltrami, *What Is Random?* (Springer, New York, 1999)
8. L. Bloomflied, *Language* (University of Chicago Press, Chicago, 1933)
9. M.A. Boden (ed.), *The Philosophy of Artificial Life* (Oxford University Press, Oxford, 1996)
10. D. Bohm, *On Creativity* (Routledge, London, 1996)
11. G. Boole, *The Mathematical Analysis of Logic* (Philosophycal Library, New York, 1948)
12. G. Boole, *An Investigation of the Laws of Thought* (Mcmillan, Cambridge, 1854)
13. B. Bosanquet, *Logic or the Morphology of Knowledge*, vol. I (Clarendon Press, Oxford, 1911)
14. B. Bosanquet, *Logic or the Morphology of Knowledge*, vol. II (Clarendon Press, Oxford, 1911)
15. M. Bunge, *Causality and Modern Science* (Dover, New York, 1979)
16. L.C. Burns, *Vagueness* (Kluwer, Dordrecht, 1991)
17. G. Cardano, *Ars Magna or the Rules of Algebra* (Dover, New York, 1993)
18. R. Carnap, *Meaning and Necessity* (The University of Chicago Press, Chicago, 1956)
19. G.J. Chaitin, *The Unknowable* (Springer, Singapore, 1999)
20. J.L. Chabert (ed.), *A History of Algorithms* (Springer, Berlin, 1999)
21. J.C. Cooley, *A Primer of Formal Logic* (Macmillan, New York, 1946)
22. D. van Dalen, *Logic and Structure* (Springer, Berlin, 1997)
23. K. van Deemter, *Not Exactly: In Praise of Vagueness* (Oxford University Press, Oxford, 2010)
24. G. Deutscher, *The Unfolding of Language* (Metropolitan, New York, 2005)
25. K. Devlin, *Logic and Information* (Cambridge University Press, Cambridge, 1991)
26. D. Dubois, H. Prade, *Théorie des Possibilités* (Masson, Paris, 1985)
27. D. Dubois, H. Prade, R.R. Yager (eds.), *Readings in Fuzzy Logic for Intelligent Systems* (Morgan Kaufmann, San Mateo, 1993)
28. M. Dummet, *The Seas of Language* (Clarendon Press, Oxford, 1993)
29. A. Feeney, E. Heit (eds.), *Inductive Reasoning* (Cambridge University Press, Cambridge, 2007)
30. P. Feyerabend, *Farewell to Reason* (Verso, London, 1987)
31. B. de Finetti, *L'invenzione della verità* (Cortina, Milan, 2006)
32. J.S. Fodor, *Representations* (The MIT Press, Cambridge, 1981)
33. G. Goldstein, *Incompleteness* (Atlas Books, New York, 2005)

E. Trillas, *On the Logos: A Naïve View on Ordinary Reasoning and Fuzzy Logic*,
Studies in Fuzziness and Soft Computing 354, DOI 10.1007/978-3-319-56053-3

34. U. Grenander, *A Calculus of Ideas. A Mathematical Study of Human Thought* (World Scientific, Singapore, 2012)
35. S.P. Gudder, *Quantum Probability* (Academic Press, Boston, 1985)
36. J. Hadamard, *Essai sur la psychologie de l'invention dans le domaine mathématique* (Gauthier-Villars, Paris, 1975)
37. H. Heyting, *Intuitionism* (North-Holland, Amsterdam)
38. J.H. Holland, K.J. Holyoak, R.E. Nisbet, P.R. Thagard, *Induction. Processes of Inference, Learning, and Discovery* (The MIT Press, Cambridge, 1986)
39. D. Hyde, *Vagueness, Logic and Ontology* (Ashgate, Aldershot, 2000)
40. M. Kac, S. Ulam, *Mathematics and Logic* (Dover, New York, 1992)
41. R. Keefe, P. Smith (eds.), *Vagueness: A Reader* (The MIT Press, Cambridge, 1999)
42. G. Lakoff, M. Johnson, *Philosophy in the Flesh* (Basic Books, New York, 1999)
43. G. Lakoff, R.E. Núñez, *Where Mathematics Comes From?* (Basic Books, New York, 2000)
44. L. Laudan, *La dynamique de la science* (Pierre Mardaga Editeur, Brussels, 1977)
45. L.M. Lederman, C.T. Hill, La simetría y la belleza del universo (Tusquets Eds., Barcelona, 2006)
46. G. Link, *Algebraic Semantics in Language and Philosophy* (CSLI Pubs, Stanford, 1998)
47. J. Mendola, *Human Thought* (Kluwer, Dordrecht, 1997)
48. K. Menger, *Morality, Decision and Social Organization: Toward a Logic of Ethics* (Reidel, Dordrecht, 1974)
49. D. Miller, *Critical Rationalism* (Open Court, Chicago, 1999)
50. M. Minsky, *The Society of Mind* (Simon and Schuster, New York, 1985)
51. S. Pinker, *The Language Instinct* (Penguin Books, London, 1994)
52. K.R. Popper, *Conocimiento Objetivo* (Editorial Tecnos, Madrid, 1974)
53. K.R. Popper, *Unended Quest* (Fontana/Collins, Glasgow, 1976)
54. K.R. Popper, *La lógica de la investigación científica* (Editorial Tecnos, Madrid, 1977)
55. K.R. Popper, *Realismo y el objetivo de la ciencia* (Editorial Tecnos, Madrid, 1983)
56. G. Priest, J.C. Beall, B. Armour-Garb (eds.), *The Law of Non-contradiction* (Clarendon Press, Oxford, 2004)
57. H. Putnam, *Reason, Truth and History* (Cambridge University Press, Cambridge, 1981)
58. W.V.O. Quine, *Pursuit of Truth* (Harvard University Press, Cambridge, 1992)
59. V.S. Ramachandran, *The Tell-Tale Brain* (Norton & Company, New York, 2011)
60. U. Ratsch, M.M. Richter, I-O Stamatescu (eds.), *Intelligence and Artificial Intelligence: An Interdisciplinary Debate* (Springer, Berlin, 2013)
61. N. Rescher, *Many-valued Logic* (Mc Graw-Hill, New York, 1969)
62. H. Reichenbach, *Experience and Prediction* (University of Chicago Press, Chicago, 1938)
63. H. Reichenbach, *Elements of Symbolic Logic* (Dover, New York, 1980)
64. H. Reichenbach, *The Theory of Probability* (University of California Press, Berkeley, 1971)
65. C. Rovelli, *La realtà non è come ci appare* (Cortina, Milan, 2014)
66. B. Russell, *An Inquiry into Meaning and Truth* (George Allen and Unwin, London, 1940)
67. B. Russell, *Introduction to Mathematical Philosophy* (George Allen and Unwin, London, 1960)
68. E. Sapir, *Language* (BiblioBazar, Charleston, 2007)
69. J.R. Searle, *Rationality in Action* (The MIT Press, Cambridge, 2001)
70. R. Sorensen, *Vagueness and Contradiction* (Oxford University Press, Oxford, 2004)
71. J.F. Sowa, *Knowledge Representation* (Course Technology, Boston, 2000)
72. E. Schröder, *Algebra der Logik* (Chelsea, New York, 1966)
73. R. Seising, E. Trillas, J. Kacprzyk (eds.), *Towards the Future of Fuzzy Logic* (Springer, Schwitzerland, 2015)
74. Y. Shoham, *Reasoning about Change* (The MIT Press, Cambridge, 1988)
75. R. Socher-Ambrosius, P. Johann, *Deduction Systems* (Springer, New York, 1997)
76. N.J.J. Smith, *Vagueness and Degrees of Truth* (Oxford University Press, Oxford, 2008)
77. R.C. Stalnaker, *Inquiry* (The MIT Press, Cambridge, 1987)

78. R.J. Sternberg, *Wisdom, Intelligence, and Creativity Synthesized* (Cambridge University Press, Cambridge, 2011)
79. J. Stuart Mill, *An Examination of Sir William Hamilton's Philosophy* (Ellisbron Classics, 2005)
80. J. Stuart Mill, *A System of Logic, Ratiocinative and Inductive* (Longmans, Green, Reader and Dyer, London, 1868)
81. P. Suppes, *Probabilistic Metaphysics* (Basil Blackwell, New York, 1985)
82. P. Thagard, *Computational Philosophy of Science* (The MIT Press, Cambridge, 1993)
83. S.N. Thomas, *Practical Reasoning in Natural Language* (Prentice-Hall, Upper Saddle River, 1973)
84. S. Toulmin, *The Philosophy of Science* (Harper & Row, New York, 1960)
85. E. Trillas, *En defensa del raonament* (Univ. de València, València, 2015)
86. E. Trillas (ed.), *Razonamiento, significado, incertidumbre y borrosidad* (UPNA, Pamplona, 2015)
87. T. Tymoczko, J. Henle, *Sweet Reason. A Field Guide to Modern Logic* (Springer, New York, 2000)
88. S. Ullmann, *Semantics* (Basil Blackwell, Oxford, 1962)
89. Z. Wang, G.J. Klir, *Fuzzy Measure Theory* (Plenum Press, New York, 1992)
90. N. Wiener, *Cybernetics* (The MIT Press, Cambridge, 1962)
91. T. Williams, *Knowledge and its Limits* (Oxford University Press, Oxford, 2000)
92. L. Wittgenstein, *Remarks on Colour* (Basil Blackwell, Oxford, 1977)
93. L. Wittgenstein, *On Certainty* (Harper Torchbook, New York, 1972)
94. A.N. Whitehead, *Symbolism. Its Meaning and Effect* (Fordham University Press, New York, 1985)
95. A.N. Whitehead, *Modes of Thought* (Free Press, New York, 1968)
96. W. Whewell, *Novum Organon Renovatum* (Parker & Son, London, 1858)
97. W. Whewell, *History of the Inductive Sciences* (Appleton and Company, New York, 1875)
98. W. Whewell, *Theory of the Scientific Method* (University of Pittsburgh, Pittsburgh, 1968)
99. L.A. Zadeh, ChA Desoer, *Linear Systems Theory: The Space State Approach* (McGraw Hill, New York, 1963)

Printed in the United States
By Bookmasters